# BURNING ICE
## ART & CLIMATE CHANGE

The Cape Farewell project was created by the artist David Buckland

**Burning Ice: Art & Climate Change**

This book documents the commitment, hard work and adventures of all those who have been part of the Cape Farewell project. Forty artists, scientists, educators and film crew have sailed into the ice of the High Arctic as part of the Cape Farewell expeditions. They have been supported by a dedicated shore-team, funders, writers and climate specialists. Climate change requires that we establish creative solutions to build a stable future.

Artwork from the Cape Farewell project features in several exhibitions, at the Bodleian Library, Oxford, December 2005, at the Natural History Museum 1 June – 3 September 2006, the Liverpool Biennial 14 September – 26 November 2006 and The Eden Project 2007/8.

**Edited by**
David Buckland, Ali MacGilp, Sion Parkinson

**Design by**
Bullet: Gorm Ashurst, Kathy Barber, Nathan Gallagher, Marije de Haas, Richard Holland
www.bulletclip.com

**Project Management**
Vicky Long

**Consultancy**
Tracey Hilty, Greg Hilty

**Published by**
Cape Farewell
239 Royal College Street
London NW1 9LT
United Kingdom
www.capefarewell.com

Cape Farewell is a registered charity 1094747.

**Printing**
Lecturis, www.lecturis.nl
Lecturis is a FSC Certified Company

**Paper**
Arctic Paper, www.arcticpaper.com
Arctic Paper is FSC Certified Company
www.fsc.org

Copyright © 2006 by Cape Farewell
All works copyright of the artists and authors. All rights reserved.

No part of this publication may be reproduced, stored in a retrieval system or transmitted, in any form or by any means, electronic, mechanical, photocopying, recording or otherwise, without the prior permission in writing from the publisher.

ISBN 0-9553109-0-3
978-0-9553109-0-4

DAVID BUCKLAND 04, 42, 94
IAN MCEWAN 10
ALEX HARTLEY 12, 158
DR. VALBORG BYFIELD 22, 74
JAMES CAMERON 28
KATE HAMPTON 28
PETER CLEGG 30
CHARLIE KRONICK 34
ANTONY GORMLEY 36
GRETEL EHRLICH 48
MAX EASTLEY 52
GAUTIER DEBLONDE 60
DR. TOM WAKEFORD 82
SIOBHAN DAVIES 86
GREG HILTY 90

GARY HUME 100
RICHARD C. SABIN 104
HEATHER ACKROYD 106, 114
DAN HARVEY 106, 114
PROF. H. J. SCHELLNHUBER 112
ANTONY GORMLEY 122
SIR DAVID KING 126
MICHELE NOACH 130
DAVID BUCKLAND 136, 154
KATHY BARBER 142
PROF. DIANA LIVERMAN 148
NICK EDWARDS 152
DR. SALEEMUL HUQ 156
RACHEL WHITEREAD 162
ROBERT MACFARLANE 170

### Cape Farewell Art and Climate Change
David Buckland, Director, Cape Farewell

Four a.m. in the morning, my bunk is no longer horizontal and I am lying against the hull of the Noorderlicht as it beats to windward 40° off vertical. As we fall off another wave there are moments of freefall weightlessness followed by four g gravity – sleep comes and dreams follow this roller coast ride for hours on end. We have just completed a 15-hour tract of oceanography measurement along the 80th parallel north to locate the Gulf Stream, or the West Spitsbergen Current, as it is called this far north. Every 50 minutes we stall the boat and lower a CTD on 100 metres of cable, which feeds directly into our onboard computers to give us detailed temperature and salinity measurements. There is an explosion of sound as the main and the schooner sails crash around in their stalled state, the seas wash onboard and we shout instructions and readings. For hour after hour we find polar waters of between 0.5°C and 1°C; ice melt and low in salinity. We are now a long way from the safe haven of the science station and fjord of Ny-Ålesund. On board are a crew of twenty artists, scientists, educators and film makers. Everyone is involved in all kinds of experiments, artworks, glacier climbing, plankton sampling and sharing information or impressions – bliss compared with this pitching, rolling and cold. All are 'enduring' below as we relentlessly keep to our task, 14 hours and still no Gulf Stream, just cold, cold, black seas.

Dr. Simon Boxall from the National Oceanography Centre is leading this chilly quest. He oversees the computers, plotting our course and measurement positions while eighteen-year old Sean, oceanographer Sarah Fletcher and I manhandle the instruments overboard, with cold wet ropes, constant dousing and crashing noise, and no other vessel for 500 miles. We are only 750 miles from the North Pole. When we halt the Noorderlicht for the fifteenth time, the satellites indicate only polar waters this far north. Simon is concerned that there is nothing to find as there is constant scientific debate as to the stability of the Gulf Stream. Suddenly 0.5°C on the surface jumps to 8°C at a 20 m depth – we have found the Gulf Stream. It had been hidden from the satellite tracking because it was flowing below the surface. I check with Simon, cocooned in his computers, and either seasickness strikes, or the information is truly shocking, because the colour drains from his face. Sean retrieves the equipment and Simon calls it a day. Three a.m. and we have completed our task and are too exhausted to comprehend the implications of our measurements. We lash down the equipment and help the Captain to gibe the boat round as we head ever further northeast, towards the top of the Spitsbergen archipelago. We retire to our bunks, which provide warmth but not calm.

It is seven years since I watched another piece of sea glide past, a sea that was calm, warm and vibrant but also carrying a message of threat. I was sitting on a terrace, it was June and we were in Venice. I had just completed photographing 'The Last Judgement', a major work by Anthony Caro in a Salt Warehouse as part of the Venice Biennale. I had been making a photographic essay for a book of the works as they unfolded over the last six months and was now tired and enjoying the watery life-style of Venice. An English newspaper I was reading carried a short report on some work done by mathematicians who had constructed a computer model of the whole North Atlantic Ocean. As a passionate sailor, with some understanding of the complexity that lies beneath the surface of the seas, I was both intrigued and incredulous. How was it mathematically possible to calculate currents, temperature differentials, temporal changes, and to do it on such a scale? More importantly, why would anyone want to do this?

Three months later I was sitting in the office of Dr. Richard Wood at the Hadley Centre, a government research centre that, amongst other things, looks at issues relating to climate. He talked to me about their extraordinary work, the HadCM3 computer model of the North Atlantic. The mathematics is beyond my understanding but there is a beauty to the structure that is almost an artwork in itself. To evaluate the model it was programmed to run from the early 1800s, as we have detailed climatic information from this point. As the HadCM3 progressed through 200 years it emulated this known data. Now, confident that it was behaving correctly, the scientists ran it into the future, feeding into it temperatures and other data to speculate on how our planet will react to anthropogenic changes. These mathematical computer models are the way scientists test hypothesis into the future. They know human actions are having an effect on the wellbeing of our planet, but they do not know if it will be lasting. Will 'nature' manage to adjust and rebalance? Are we causing permanent damage to our planet? Are we heading towards catastrophic scenarios? Richard introduced me to colleagues at the National Oceanography Centre and in turn they put me in touch with other climate scientists working in the U.S.A. and worldwide. This model of the ocean has led me to investigate the troubling changes happening to the Earth and given me a crash course in climate awareness. Vitally, I also came to realise how closely the climate science community worldwide has been working for many years and just how concerned they are for the planet and our future.

I felt there was a need to find a different way of communicating their important message as the scientific delivery was being ignored. I hatched a plan. I would ask artists, because they are our most creative communicators, to join a group of scientists and educators to sail north on an extraordinary expedition to the front line of climate change. There they would witness, understand and react to the rapidly changing Arctic environment. I walked into the offices of the National Endowment of Science, Technology and Art (NESTA) in London, with a chart under my arm and a story to tell, and they listened, understood, and provided me with an investment of seed money to work up a robust proposal. And so it began.

My studio on top of our family house in Camden became Cape Farewell HQ. Emma Gladstone and I started to build a project on the strict understanding that Emma would not join our gruelling sailing passage north as she was once famously seasick on a pedalo crossing the Bay of Naples. The original plan was to sail from Scotland to Cape Farewell in Southern Greenland but in fact we decided to sail much further north to where the Gulf Stream 'sinks' off the coast of Spitsbergen. A life-long friend, Arctic traveller and writer Gretel Ehrlich, put us in contact with the 100-year-old schooner the Noorderlicht. This vessel was perfect for our needs, she had the capability to accommodate our crew of 20 and was just stable enough to conduct oceanography experiments from. Crucially, her captain also knew the waters of the Spitsbergen archipeligo. The adventure evolved and this book is a document of the three Cape Farewell expeditions. The first commenced in Tromsø in northern Norway, sailing north in May 2003 to explore the eastern coast of the Spitsbergen islands. May is a time of 24-hour daylight when these islands are just becoming free of Arctic ice. On the second expedition in September 2004 we joined the Noorderlicht in Longyearbyen and for 20 days circumnavigated the archipelago through the Hinlopen Streetet before heading south to Tromsø. It was a time of autumn equinox and Northern Lights. For the third expedition we flew to Longyearbyen and on snow scooters travelled 60 km over mountains and glacial valleys to join the Noorderlicht locked in ice in the Tempelfjorden. It was March 2005, the temperature -35°C, daylight and night were of equal length and the Northern Lights showed their magic each night. These expeditions were the icing on the cake; during the times in between the Cape Farewell HQ team worked hard. Alex Lambert managed the project after Emma and joined the second expedition and now Vicky Long holds this complex endeavour together. The Arts Council became our funding partner, enabling the artists to create bodies of work.

On the first expedition we gathered together the twenty-strong team, most of whom had not previously met, at Heathrow airport at 5.30 in the morning. We had 600 kg of excess baggage; film equipment, oceanography machinery, artists' tools and arctic survival gear. For everyone, this was an active leap of faith. We finally boarded the Noorderlicht in Tromsø 7 hours later and astonished Maaike, the ship's first mate, with our amount of equipment and the diversity of our crew; including eight nationalities and ranging in age from 11 to 59. Ted, the captain, eventually emerged from his cabin announcing that we should sail immediately to catch a favourable weather system; little did we know the implications of this decision! Sailing into the 24-hour daylight of the Arctic was why we had come, so let's get on with it, we thought. Enchanted by the Noorderlicht, we hadn't counted on seven sails each taking five people to hoist, a four-hour watch system and force 6-7 winds on the nose. We sailed the Devil's Dance Floor, for two very rough days and day-lit nights. Only four of us escaped sea-sickness but most reported for duty and I expected to be lynched every time I came on deck – why have you brought us here and is there 18 days of this? I promised we would shelter for the night behind Bear Island, but we arrived to a kilometre of ice offshore, no landing was possible. A lynching was averted, however, by the sighting of a polar bear swimming towards the ship and the Arctic magic started. North we sailed along the ice pack, the seas calmed and the light was beyond imagination. It is well documented that Arctic travellers get high from the air and light, and we embarked on two weeks of euphoria. We crept and forced our way through the ice into the fjord of Hamsund, where we heard the very welcome sound of the anchor chain and merriment commenced. Anna, our cook provided a feast, wine flowed on very empty stomachs and sound artist Max Eastley lowered hydrophones into the seas to hear the singing of bearded seals as they mated, a calling sound of perfect pitch, travelling tens of miles through the sea. The boat, being made of steel, amplified these sounds and we went to our bunks blissfully serenaded by the seal song from the other side of the hull.

The days following were a heady mix of activity; long walks over ice fields and mountains, geological observation, filming interviews, and scientific research. Dr. Valborg Byfield was enthusiastically engaged in plankton trawling during the expedition. We all controlled our own projects but relied on each other's help in their realisation. Any shore activity was carefully planned not to harm the beautiful environment and not to be eaten by polar bears. There is no defence against these incredibly powerful animals. They rarely see humans, can outrun you, out-swim you and are dangerous, if hungry. A gun is the only defence, and every shore party carried one, although shooting a bear is the very last thing you would want to do. The melting ice cap is already depriving these animals of their hunting grounds, and it is unlikely that they will survive in the wild beyond a hundred years from now.

The eight artists onboard wrestled with the beauty and scale of the place and how to make sense of it all. Each was a skilled practitioner in their chosen medium: Gary Hume, painter; Gretel Ehrlich, writer; Gautier Deblonde, photographer; Dan Harvey, sculptor; David Hinton, film director; David Buckland, lens-based artist; Nick Edwards, 16 mm cameraman; Max Eastley, sound sculptor; Suba Subramaniam, Bharata Natyam dancer and science teacher. The second expedition included Heather Ackroyd, sculptor; Alex Hartley, photographer and sculptor; Michèle Noach, Artoonist and Kathy Barber, web and new media designer. The final expedition, when the Noorderlicht was locked in ice included: Antony Gormley, sculptor; Rachel Whiteread, sculptor; Siobhan Davies, choreographer; Ian McEwan, writer.

The Cape Farewell invitation was open: come to the Arctic, engage with scientists and, we hope, be inspired to make art. Art is never a guarantee; many a time I have followed a path of enquiry only to abandon it in frustration. What is extraordinary is that each artist has produced work that celebrates authorship. When unpacked, each artist has in some form responded to this cold, Arctic place and the way it is changing as the Earth warms. They have all told the story on a human, rather than a planetary, scale.

The number of art disciplines represented onboard led to a curious range of activity. Alex Hartley initially named a glacier wall climb he made. Not satisfied with this, he then set out to discover an island with the intention of going ashore to 'conquer' and name it. The island he found is only two years old and came into existence because of glacier retreat. The writers wrote, contextualising the climate debate and how it is manifest in these Arctic deserts. Dan and Heather endlessly started 'drawings', outcome unknown. Their days were spent immersed in the physical nature of this Arctic world; testing the sea and the cold, casting ice, examining objects eroded by weather and dyes made of glacial mud. They came to know this place and it formed their art works. This Arctic world we were in was changing, human behaviour was changing it, and this reality was embedded in Dan and Heather's work. Max Eastley showed us how to see using sound, to experience and hear nature in fast-mode. We all left a trace there, but our footprints are gone now, absorbed in the endless movement of ice and weather. The artists' Arctic works are documented in this book in photographs and in their writings. Working from our urban studios, many of us continue to make artworks inspired by the Arctic expeditions and our changing world. These and the art from the expeditions will be on public view in a series of exhibitions throughout the UK and abroad. The Cape Farewell film, directed by David Hinton, shows the physical reality of working in the Arctic; the sea, the ice, the limitations of living bodies in a very cold environment. At -35°C with wind chill on top, cables snap, film solidifies, frostbite and sea-sickness attack and bones are bruised. The documentary film is a testament to those that made it, as much as to those in it: 2,500 nautical miles, mostly sailed, 200 hours of film shot, much debate and laughter.

Education became central to our thinking and, in partnership with the Geographical Association, we developed 'Extreme Environments', a new module to accompany the Geography 21 GCSE. This innovative module, funded by NESTA was included in a film directed by Colin Izod of education production company Big Heart Media, an interactive CD-rom, a teachers' manual written by Fred Martin and a student resource of photographs from the expedition. This was a proud moment for the Cape Farewell team as it is so difficult to get new teaching programmes accepted within the curriculum. It was the first time climate change had been taught in schools as part of the syllabus for sixteen-year olds. We have subsequently followed this with a science module, developed in partnership with the Nuffield Foundation, National Oceanography Centre and Big Heart Media, based on the work done during the Cape Farewell expeditions. It has already been introduced as part of the new Science 21 GCSE. Again this brought the discussion of the complexities of climate change to schools, informing the very people who stand to be most affected. It is the younger generation who will have to pick up the tab for our wasteful and excessive use of energy. We continue to build educational material from the Cape Farewell archive. Plans have now moved towards using the work done by the artists on these expeditions in schools too.

To borrow an idea from Ian McEwan, we are asking today's generation and our world leaders to act now to avert future calamity. This is unprecedented in human history and runs counter to our nature. As McEwan points out: 'We've got ourselves into the situation where we are having to address the needs of people unborn. And not our own children. Maybe not even our children's children but generations after that. Even the most idealistic of thinkers and actors on the world stage in the past have only really addressed themselves to problems in the present.'

Walk outside now and you might not detect any obvious indication that things are altering because of climate change. There is, however, a very direct message from the scientific community that we have to act now to mediate the effect our lives are having on the wellbeing of our planet. If we wait until it is irrefutably clear from day-to-day observations that climate change is real before we implement the necessary reduction of green house gas emissions, it will already be too late. In this book I have asked some world-renowned climate scientists and thinkers to explain their concerns in detail. They have responded and provided lucid essays that collectively give a very clear position. This should encourage us all to become involved in the evolution of a more sustainable future. Seven years ago, when the Cape Farewell enterprise was initiated, there was very little public debate on these issues, thankfully we are now much more climate-aware and willing to engage with the necessary changes.

Unfortunately climate change is with us into the future, $CO_2$ is a very stable gas and there is no quick solution. Any positive changes we implement will only come to fruition many years from now. The last seven years have been very exciting and if this project is achieving something positive in the field of climate change, then I can look optimistically towards our future. The human behaviour that has caused climate change has only been with us for the 200 years since the start of the Industrial Revolution. Aren't we due another turning point, towards a less selfish, more symbiotic time?

8 David Buckland

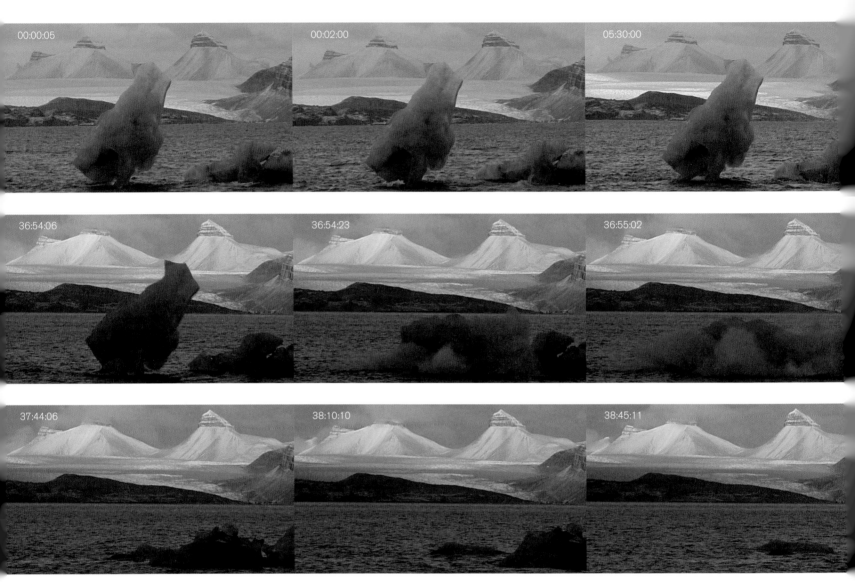

⊕ **Sinking Ice** 2004 / 78°55N, 12°00E

'IN 30 YEARS WE HAVE ACTUALLY GOT TO CHANGE OUR LIFESTYLES. I DON'T KNOW IF HUMAN BEINGS HAVE THE CAPABILITY FOR THE KIND OF CHANGE THAT IS NECESSARY. YET, MAYBE THERE IS A CHANCE. IT IS DOWN TO THE MESSENGER, THE NARRATIVE, THE STORY, TO MAKE CHANGE POSSIBLE.'

**The Hot Breath of Our Civilisation** Ian McEwan, Novelist

THE PRESSURE OF OUR NUMBERS, THE ABUNDANCE OF OUR INVENTIONS, THE BLIND FORCES OF OUR DESIRES AND NEEDS ARE GENERATING A HEAT – THE HOT BREATH OF OUR CIVILISATION. HOW CAN WE BEGIN TO RESTRAIN OURSELVES? WE RESEMBLE A SUCCESSFUL LICHEN, A RAVAGING BLOOM OF ALGAE, A MOULD ENVELOPING A FRUIT.

WE ARE FOULING OUR NEST, AND WE KNOW WE MUST ACT DECISIVELY, AGAINST OUR IMMEDIATE INCLINATIONS. BUT CAN WE AGREE AMONG OURSELVES?

WE ARE A CLEVER BUT QUARRELSOME SPECIES – IN OUR PUBLIC DEBATES WE CAN SOUND LIKE A ROOKERY IN FULL THROAT. WE ARE SUPERSTITIOUS, HIERARCHICAL AND SELF-INTERESTED, JUST WHEN THE MOMENT REQUIRES US TO BE RATIONAL, EVEN-HANDED AND ALTRUISTIC.

WE ARE SHAPED BY OUR HISTORY AND BIOLOGY TO FRAME OUR PLANS WITHIN THE SHORT TERM, WITHIN THE SCALE OF A SINGLE LIFETIME. NOW WE ARE ASKED TO ADDRESS THE WELL-BEING OF UNBORN INDIVIDUALS WE WILL NEVER MEET AND WHO, CONTRARY TO THE USUAL TERMS OF HUMAN INTERACTION, WILL NOT BE RETURNING THE FAVOUR.

PESSIMISM IS INTELLECTUALLY DELICIOUS, EVEN THRILLING, BUT THE MATTER BEFORE US IS TOO SERIOUS FOR MERE SELF-PLEASURING. ON OUR SIDE WE HAVE OUR RATIONALITY, WHICH FINDS ITS HIGHEST EXPRESSION AND FORMALISATION IN GOOD SCIENCE. AND WE HAVE A TALENT FOR WORKING TOGETHER – WHEN IT SUITS US.

ARE WE AT THE BEGINNING OF AN UNPRECEDENTED ERA OF INTERNATIONAL CO-OPERATION, OR ARE WE LIVING IN AN EDWARDIAN SUMMER OF RECKLESS DENIAL? IS THIS THE BEGINNING, OR THE BEGINNING OF THE END?

# 'OUR GOAL HAD BEEN TO FIND A NEW ISLAND THAT HAD BEEN REVEALED BY THE RETREATING GLACIERS AND TO CLAIM IT FOR OUR OWN.'

**Nymark (Undiscovered Island)** Alex Hartley, Artist

In the early evening of the 19th day of September, year of our Lord two thousand and four, we passed through the fast flowing narrow sound of Heleysundet. This tight passage separates the main island of Spitsbergen from its smaller neighbour Barentsøya. According to the pattern established over the last several days, we continued scouring the coastline for our ever-elusive quarry.

Suddenly, on rounding the southern point of Bakanbukta, the cry went out … maps were checked, the compass consulted and a small expeditionary group was hastily assembled and subsequently dispatched in the support craft.

Our small vessel headed towards the glacial edge of Sonklarbreen still unsure that our days of searching could have been so finely rewarded. Our goal had been to find a new island that had been revealed by the retreating glaciers and to claim it for our own. The island towards which we now sped exceeded all expectations. Roughly an hundred paces long, fifty wide, and fully thirty feet high. The terrain was largely muddy moraine holding many varieties of rock within its frozen clutches. We came across the nests of eider ducks, and several purple sandpipers were seen quickly departing.

Our new island was surrounded by a beach on all sides, and could be circumnavigated on foot in roughly ten minutes. The geography of its interior was marvelously varied, with towering mountains (some higher than twenty feet). There were valleys and even a small frozen lake (a member of our group unkindly likened this to a pond, but little notice was taken of his comments).

We surveyed the island taking longitude and latitude readings for all features and extremities. A cairn was built and in the age old style, a claim note was placed inside a tin-can and this in turn was inserted into the cairn. The note stated, in both English and Norwegian, notice of our claim on the newly revealed land. Upon our return to the mainland, our new island will be charted and I will submit it for inclusion in all subsequent maps. The land will be named and registered. The name has not yet been finalised, but I feel the most obvious, Alex Hartley Land, may cause some ill feelings amongst my fellow crew members.

Nothing has yet been ruled out; annexation, independence, tax haven, wild life sanctuary, short let holiday homes or time shares. Postcards will be printed and a major architectural competition will be launched. Engineers will be consulted as to how best to keep all the mud together and prevent any shrinkage of our island.

On the morning of the twentieth we traversed the glacial edge in the Noorderlicht coming within an arm's distance of the towering blue face. Then, after breakfast, we turned our backs on our newly discovered territory and set sail for points south. It was with a heavy heart and a tear in my eye that I watched it disappear. This land so newly revealed, land which had lain below the crushing weight of the ice for thousands of years, land on which no human had ever stood. This new land, so freshly released, was indeed our land, and part of me was left behind there.

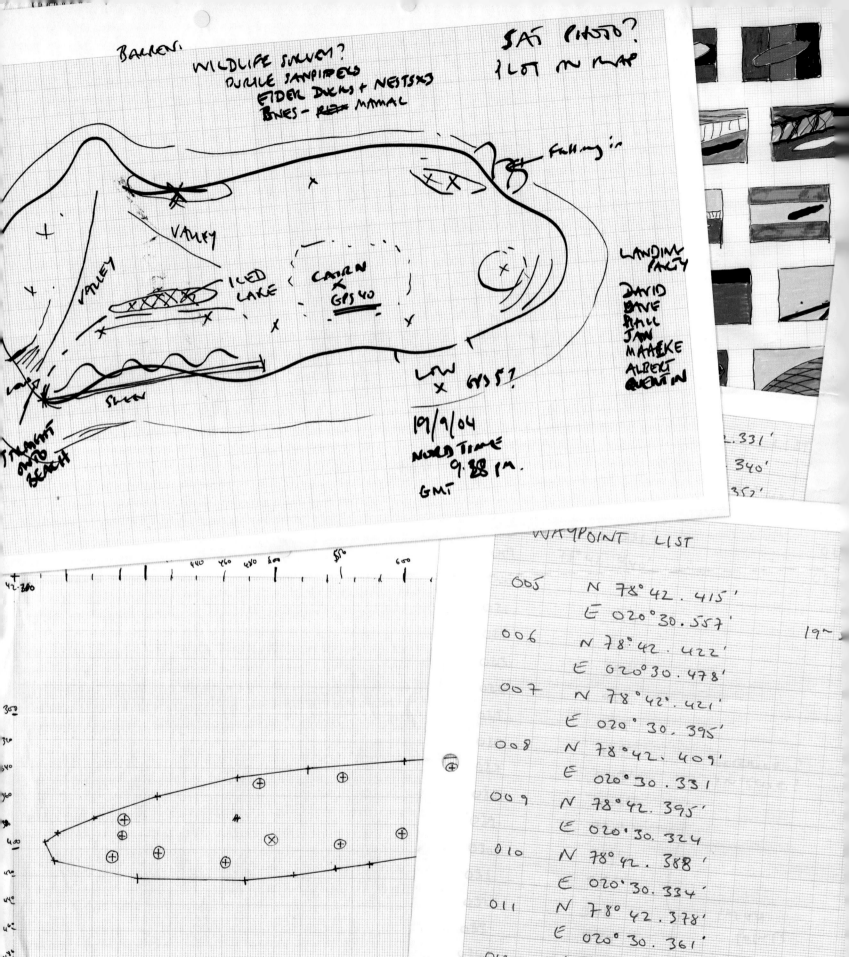

**BERGVESENET**
MED BERGMESTEREN FOR SVALBARD

Advokatfirmaet Schjødt AS
Postboks 2444 Solli

0201 OSLO

MOTTATT
2 9 JUL 2005
Besvart/Sign:

Deres ref.:

Vår ref.: BV utg. 0688/05 PZB/ES
Sak nr. BK/05

Dato: 27.07.2005

ANMELDELSE AV FUNNPUNKT PÅ SVALBARD

Det vises til Deres oversendelse av 22.07.05.

De ber i Deres oversendelse om at bergmesteren avbryter 10 måneders fristen slik at ny frist kan fastsettes for å avhjelpe manglene i henhold til Bergverks-ordningen §9 nr.4.

Det er dessverre flere mangler ved den oversendte anmeldelsen, i tillegg til at anmelderen ikke har den nødvendige tillatelsen for å kunne anmelde funnpunkt. Sistnevnte fremgår av §9 nr.1 hvor det heter at: "Opdager nogen ved lovlig søkning..." så kan funnet anmeldes. For at søking etter naturlige forekomster av de i §2 nevnte mineraler og bergarter kan foretas på Svalbard må man ha søkeseddel fra bergmesteren. Undertegnede kan ikke se at Bergvesenet med Bergmesteren for Svalbard har utstedt søkeseddel til hr. Hartley. Anmeldelsen oppfyller således ikke kravet om lovlig søkning i §9 nr.1 og er derfor ikke gyldig.

Avslutningsvis bør det kommenteres at under anmeldelsens pkt.8 Funnets art angis "unidentified mineral/rock". Dette oppfyller ikke kravet om "...erhverve og utnytte naturlige forekomster av kull, jordoljer eller andre mineraler og bergarter som utvinnes gjennom bergverksdrift...", jf. §2 nr.1. Her ber vi Dem spesielt iaktta at funnet må antas å kunne utnyttes gjennom bergverksdrift. Dette tilsier at man vet hvilket mineral eller hvilken bergart man har funnet slik at man kan foreta en vurdering av en mulig bergverksdrift.

Med hilsen

Per Zakken Brekke
bergmester

Leiv Eirikssons vei 39 - Postboks 3021 Lade, N-7441 Trondheim - Telefon +47 73 90 40 50 - Telefax +47 73 92 14 80
E-post: mail@bergvesenet.no - Giro: 7694.05.05883
Svalbardkontor: Telefon +47 79 02 12 92 - Telefax +47 79 02 14 24

**Nymark (Undiscovered Island)** 2004 / 78°55N, 19°05E

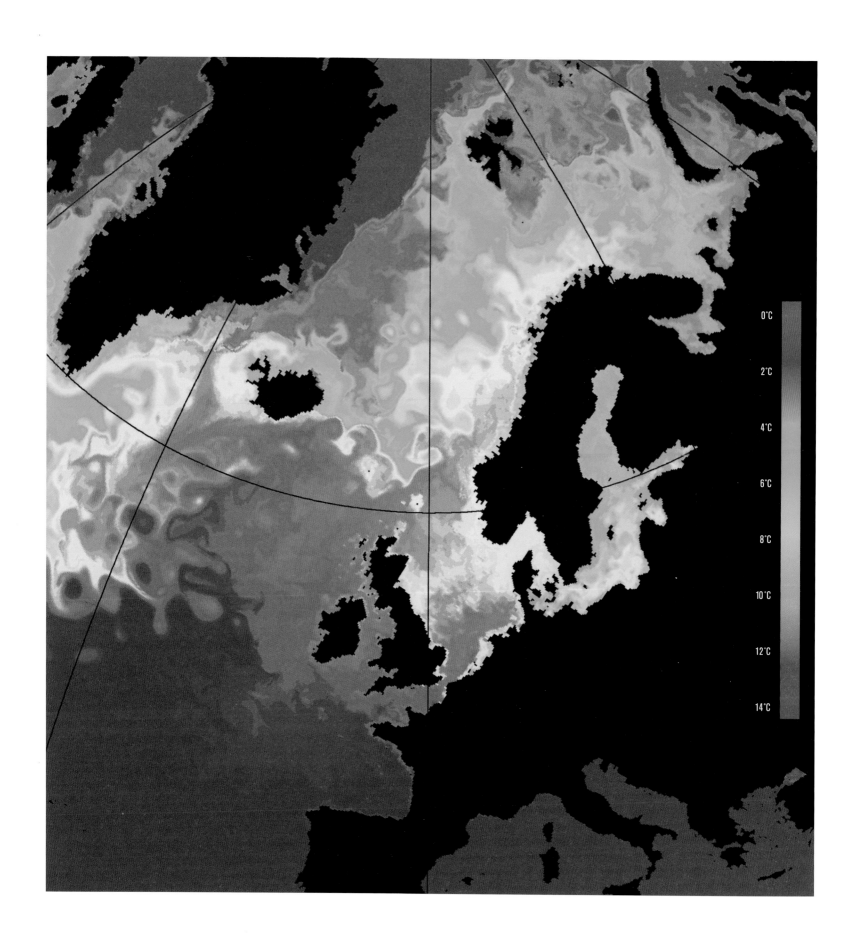

## The Atlantic Conveyor
Dr. Valborg Byfield, Oceanographer, National Oceanography Centre, Southampton

Maps of sea surface temperature show clearly how warm water travels north towards Svalbard. In this map – based on output from OCCAM, the NOC's ocean circulation model – Gulf Stream water is coloured red through yellow to green, and water of Arctic origin is blue and purple. Ocean circulation models such as this are a key tool to help us understand the role the ocean plays in our climate system.
NOC

The ocean helps shape climate around the world. Ocean currents transport heat from the equator towards the poles, releasing heat into the atmosphere, and also influencing regional rainfall patterns.

Ocean circulation, a global system of surface currents and deep currents, is powered by two different 'engines'. Movement in the top few hundred to a thousand metres is driven mainly by prevailing winds. Vertical circulation is driven by cold, salty water sinking at high latitudes, returning towards the equator at depth and being replaced by warm water moving towards the poles at the surface. This is known as the thermohaline circulation (or THC) from the combination of temperature (thermo) and saltiness (haline) that controls high-latitude sinking. The flow we can measure is a combination of both systems; wind-driven circulation and the THC.

### Why the Atlantic is special
The Atlantic is the only ocean where heat is transported north across the equator. Here warm surface water from the tropics reaches further north than anywhere else. Relatively warm, salty water from the Gulf Stream system remains at the ocean surface west of Svalbard to a latitude of about 80° before it dips underneath the much fresher and less dense polar water. The heat released by this warm water makes the climate in the countries of northwest Europe warmer than at similar latitudes elsewhere. The results of this warm flow can also be seen in the extent of Arctic Sea ice, which differs markedly from that in the Pacific region of the Arctic. The effect of this Atlantic heat conveyor is most noticeable in autumn and winter.

The relative warmth of the northern North Atlantic is due to the unique role this ocean plays in thermohaline circulation. Cold, dense water sinking in the northern North Atlantic drives the Northern Hemisphere loop of the global THC. In the Pacific there is no such area where salty subtropical water can travel far enough north and cool down sufficiently to sink. Climate scientists estimate that the Atlantic heat conveyor is responsible for up to a quarter of all heat transported from the equator towards the poles.
A slow-down in the North Atlantic loop of the THC may therefore have consequences not just for the North Atlantic region, but for the entire globe.

### How could the THC slow down?
The North Atlantic THC is controlled by the sinking of dense (cold and salty) water at high latitudes. The density of seawater is a result of both temperature and salinity (salty water is denser than fresh water, and cold water denser than warm water). Although the Gulf Stream water is saltier than the deep water below, it is much warmer, so its density is lower, and it remains on the surface. On its journey north, the water releases heat to the atmosphere, and cools gradually, until sinking can begin. This takes place in the Nordic Seas (between Greenland, Iceland and Norway with Svalbard), and is known among oceanographers as North Atlantic Deep Water (NADW) formation. NADW is also formed west of Greenland, in the Labrador Sea.

The surface water is still warmer than the deep water, but it also saltier, so its density matches that of the deeper water, allowing the two layers to mix. Should the surface water freshen for some reason, it would have to cool further before it can sink. Sufficient freshwater input might reduce salinity to the extent that the surface water could not possibly sink, even at sub-zero temperatures.

Paradoxically, global warming could have precisely this effect. Increased rainfall and the melting of sea ice, glaciers and the Greenland ice sheet are all possible consequences of higher temperatures that could potentially reduce North Atlantic surface salinity sufficiently to slow down, or even stop, the formation of deep water. If this happens the consequences could be a colder winter and a drier climate for northwest Europe, and changes to weather patterns in other parts of the world.

We do not yet know how likely a slow-down is, how soon it might occur, or what the detailed consequences would be for climate in different parts of the world. Finding out is one of the major challenges facing current climate research.

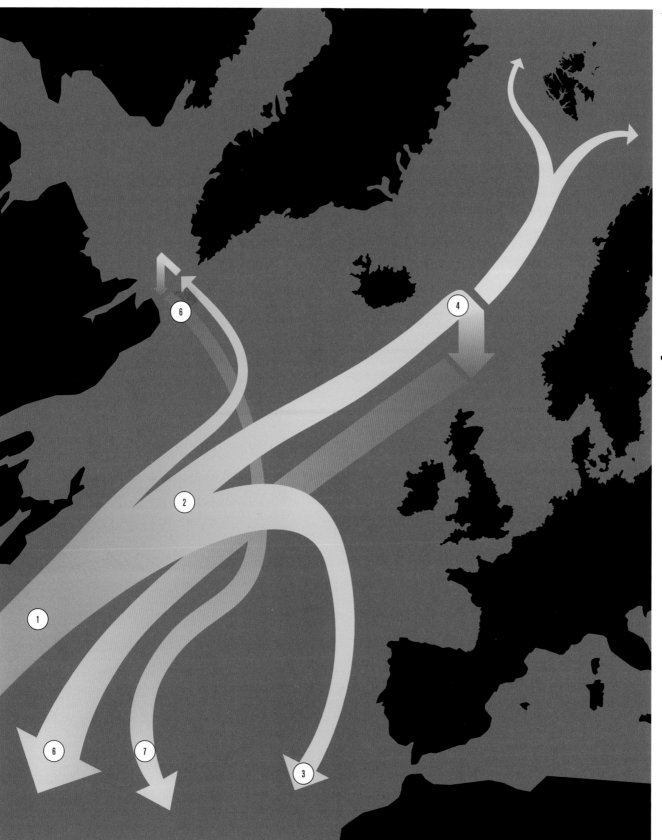

◀ **North Atlantic Conveyor**

**1** In the upper layer of the ocean warm water from the tropics travels north in the Gulf Stream.
**2** This water crosses the Atlantic towards Europe in a slow, broad current; the North Atlantic Drift.
**3** Some of this water heads back south past West Africa as the Canary Current.
**4** Some continues north, loosing heat to the air above. In the Nordic Seas it becomes cold and dense enough to sink.
**5** This cold, dense water is the North Atlantic Deep Water, this flows south at a 3-5000m depth.
**6** Another branch of the Gulf Stream water loops back west where it, too, cools and sinks.
**7** This returns south above the water from the Nordic Seas. NOC/A. Coward

▶ **The Nordic Seas Engine**

**1** North of the UK the Gulf Stream water continues towards the Arctic as the Norwegian Current
**2** Through the Nordic Seas the salty tropical water mixes with colder Arctic water and sinks.
**3/4/5** This cold, dense, deep water flows into the main Atlantic basin by three routes; over half (some 60%) through a deep channel near the Faroe Bank, most of the rest flows through the Denmark Strait between Greenland and Iceland and some occasionally crosses the Iceland-Faroe ridge.
**6** These flows merge and head south as North Atlantic Deep Water.
**7** On the surface, the warm current continues north, mixing with polar water, gradually diminishing. North of the Arctic Circle the core splits.
**8** One branch flows east into the Barents Sea.
**9** The western branch continues past Svalbard as the West Spitsbergen Current.
**10** Warmer, saltier Gulf Stream water remains near the surface along the Spitsbergen coast, but eventually dips under the colder, fresher polar water before it flows into the Arctic Ocean northwest of Svalbard. NOC

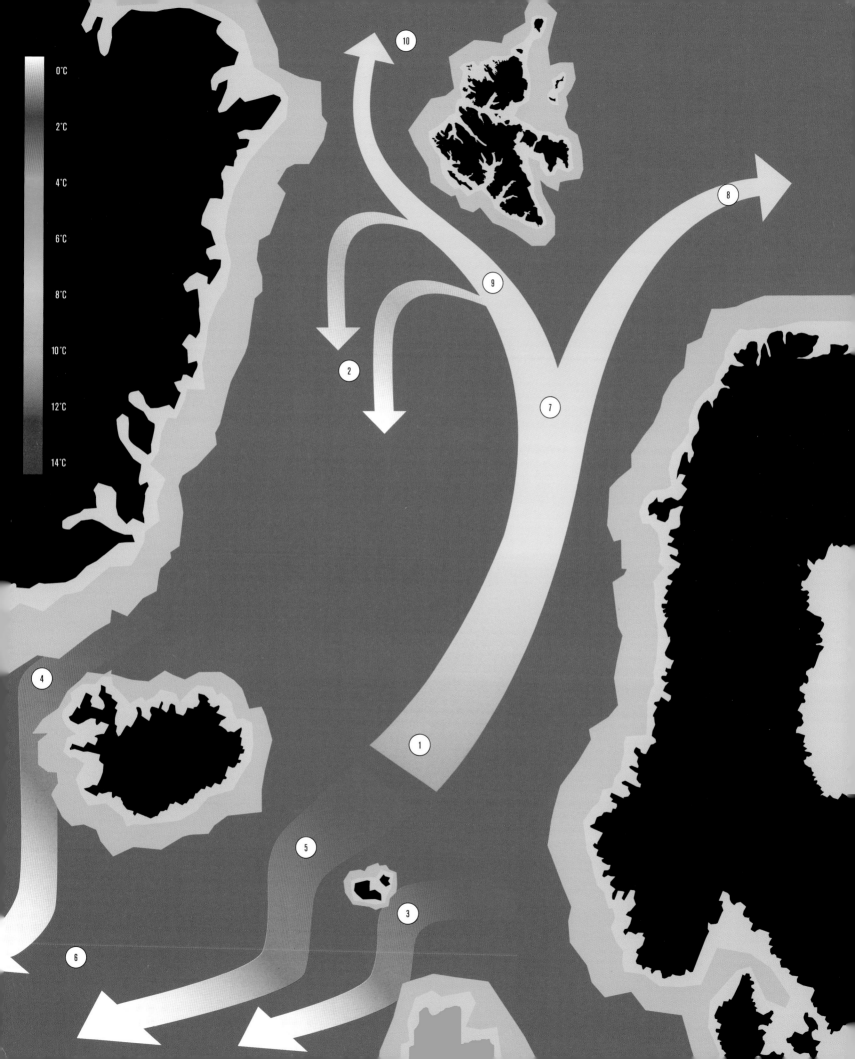

**Journal 14/09/2004** Dr. Simon Boxall, Oceanographer, National Oceanography Centre, Southampton

Today was the day that the science team were in the spotlight. After an illuminating experience the night before seeing the German atmospheric laser probe the Arctic skies to measure cloud and gas content at Ny-Ålesund, we switched from hi-tech to low with Mike's tin can ice corer. After some chilled hours with Emily trying to take a core to investigate the structure of the ice we realised why professional corers are diamond tipped – 5,000 year old ice is very hard.

Fortunately scienctific credibility was saved when we managed to predict where the drifting buoy supporting Dan's developing plaster sculpture had gone. He suspended a block of plaster beneath a float for 2 days which we then let drift around the fjord – to see how nature would sculpt it. The results were artistically stunning and scientifically interesting – this could be the birth of a new analytical tool for future use.

Then the big event – the transect from Ny-Ålesund due west along the 79 degree parallel to try to locate the remnants of Gulf Stream water – better known to us as the Norwegian Atlantic current.

This was the point at which the scientist's popularity with the artists was put to the test as we pushed out into the heavy swell of the Arctic Sea.

We sampled at 10 points heading west into the open ocean, heading to the point where the depth of water reached 1,000m – the shelf edge. Here we found surface Arctic waters – fresh and cold (2°C) from the glacier melt waters.

At only 30m depth were the waters that once flowed past the hot Florida coast – the warm salty waters of the Norwegian Atlantic current (over 7°C).

At the bottom of the ocean we found the very cold (below 0°C) waters of the Arctic Sea.

Our conclusion? Difficult from this 'snap shot' but let's just say the Atlantic current, which maintains the ice-free waters off to the west of Svalbard, is hotter now than most of the text books of previous years indicate!

**Banking on a cleaner future** James Cameron & Kate Hampton, Climate Change Capital

Most people care about the planet. But the problems seem so great and the power of the individual so small. This is especially true of climate change, caused by emissions of greenhouse gases into the atmosphere, which are the result of almost every kind of economic activity, from heating and cooling homes and offices, to producing electricity and fuelling transport. Can we really change the way the whole world works?

Imagine if labelling on food included not just calories and sell-by dates but told us about the carbon emissions associated with transporting it? If products were priced to reflect the damage, we might think twice about buying green beans from Kenya instead of Kent. Wouldn't you wait to turn on the kettle if a smart electric meter told you that power plants were burning coal for the next half hour instead of cleaner gas, and that this would both cause pollution and cost you more? This applies to other environmental issues too. Would you buy roses that came with information about the use of pesticides during cultivation, pesticides that threaten the health of local communities in places like Colombia? Decisions we take everyday have global impacts but with information and price signals telling us what hurts the planet, thinking about our global impact could be empowering rather than depressing.

Climate change is essentially an investment problem. Investors need to understand the costs associated with dirty investments and the opportunities associated with clean ones. In 2005, governments introduced a maximum limit, or 'cap', on the emissions of every large energy-intensive installation and power plant in the European Union. This artificial shortage created by government policy means that emissions from those sectors, measured in tonnes of carbon dioxide, now have a value and therefore a price. The price signal means that every time industry wants to do something that involves emitting carbon dioxide, such as making steel or electricity, they will have an incentive to think about how to do this while emitting less carbon dioxide in order to avoid paying the carbon price. Those that reduce their emissions by more than their cap are allowed to sell the difference to companies that have not. This is called emissions trading because companies buy and sell units representing tonnes of emissions of carbon dioxide. These units are bits of electronic data, monitored by the government to ensure that they represent real emissions.

Emissions trading allows companies to choose the cheapest way to cut emissions. This might involve buying spare tonnes of carbon dioxide from a factory in another European country or even outside the EU. The Kyoto Protocol, the international treaty that has started to tackle the climate change problem, includes mechanisms to identify and verify projects that reduce emissions all over the world. Under Kyoto, a company in the UK can invest in, say, China or Russia, where it might be cheaper to reduce emissions than at home, and use the tonnes it gets to meet its obligations under EU law. The atmosphere does not care where emissions reductions happen, so the Kyoto system helps companies find the cheapest emission reductions first, making it easier to tackle climate change.

In 2001, President Bush pulled out of the Kyoto Protocol. The EU decided to move ahead regardless, bringing the rest of the world (except Australia) with it. The Kyoto agreement provides emissions limits for industrialised countries up until 2012. Despite growing support for action in the Senate and at State level in California and the North East, the lack of US participation remains an impediment to real international action. Even countries that have signed up to Kyoto targets are slow to implement the policies that will deliver emissions reductions.

In order to raise capital to fund the solutions to climate change, investors need to know that they will get a financial return on their investment. In the energy sector, this means being able to estimate risks and rewards for 15 years. Potential changes in government policy represent a major risk, so reducing this risk by providing long-term

# 'IMAGINE IF LABELLING ON FOOD INCLUDED NOT JUST CALORIES AND SELL-BY DATES BUT TOLD US ABOUT THE CARBON EMISSIONS ASSOCIATED WITH TRANSPORTING IT?'

regulatory visibility encourages investment. But the provision of regulatory visibility requires political leadership. In the absence of US involvement at federal level, this leadership must come from the EU.

In March 2005, European Council conclusions identified an increase in average global temperature of 2°C above pre-industrial levels as the threshold above which unacceptable risks would occur. To avoid warming the planet above this threshold may, or may not, be possible, given current levels of emissions, but if it is to be done, it requires urgent action. Investment decisions taken between now and 2015, the middle of the next Kyoto period, will be essential in determining success or failure in achieving this objective, or indeed less ambitious climate goals.

Significant amounts of Europe's carbon-intensive plants will reach the end of their lives during the next decade and will have to be replaced, providing an opportunity to lock-in low carbon power generation. Meanwhile, emerging economies, especially China and India, are investing in large amounts of new electricity capacity. Consequently, international capital cycles and infrastructure replacement are becoming synchronised, presenting global opportunities for economies of scale in clean technology.

While an international agreement governing emissions after 2012 is negotiated, EU governments will have to commit to emissions reductions in advance of securing similar commitments from competitor countries. Negotiations have already started but are likely to last until the end of the decade. The current perception is that this kind of leadership presents costs, at least in the short term, either by burdening the taxpayer through subsidies, or by damaging the international competitiveness of some energy-intensive industries, or by pre-empting countries' negotiating positions. These costs are not yet perceived by government to be outweighed by benefits, such as establishing a clear first-mover advantage in clean energy markets.

However, not to extend the regulatory framework would send a strong negative signal to the private sector and developing countries. Any break in the regulatory framework would make it much harder to subsequently mobilise capital for emissions reductions and to get a post-2012 agreement. Having taken the necessary step of introducing the Emissions Trading Scheme (ETS), the EU now has to ensure that other countries engage in the climate effort in order for the EU to realise a first-mover advantage. The EU needs to succeed in reducing emissions in line with Kyoto targets while protecting European competitiveness, in order for others to identify the EU example as one to follow.

Last year saw unprecedented diplomatic outreach on climate change during the UK's G8 and EU presidencies and the launch of post-2012 negotiations in Montreal. Following the entry into force of the Kyoto Protocol, investment in emissions mitigation in developing countries has accelerated through the Clean Development Mechanism. Political space has further expanded in response to rising energy prices and growing scientific evidence of the severity of climate change impacts. Developing countries are currently more engaged than ever. It is essential the EU capitalises on this diplomatic and economic window of opportunity before it closes. Such an effort would prepare the ground for successful negotiations regarding the post-2012 climate regime and make a significant contribution to establishing a new North-South consensus on energy security and sustainable development.

Population and consumption trends suggest that, in the future, countries and companies that create more wealth using fewer resources will do best. Getting more from less should be something that the EU gets good at, leading a global transformation that will improve not only our own productivity, but the welfare of other countries and future generations. It is also something that we, as individuals, need to take seriously.

**Journal 09/04/2005** Peter Clegg, Architect

How do we envisage global warming? Do we think about a parched English landscape with dying beech trees? Redundant ski resorts, or continuous disastrous floods in Bangladesh? The first major ecological changes are likely to occur in polar regions, and with the shrinking of the Arctic ice cap, the islands of Svalbard are likely to experience dramatic ecological changes which will result in, amongst other things, the loss of the habitat for the polar bear that proudly occupies the top of the food chain. The group of artists and scientists on the Cape Farewell trip spent 5 days there last week, at what seemed to be the very edge of the world.

And how do we envision the causes of global warming? We know that the major culprit is manmade carbon dioxide emissions and we are becoming aware of the concept of a kilogram of carbon dioxide as a measurement of global pollution from cars and buildings. But what do we understand by a kilogram of $CO_2$? How can our minds grasp the weight of a gas? We understand a gallon of petrol, a pint of beer, a pound of sugar more because we see them as volumes than feel them as weight. Some time ago it occurred to me that it might be helpful to try to define the kilogram of $CO_2$ as a space rather than mass. One kilogram of $CO_2$ at atmospheric pressure occupies 0.54 of a cubic metre. That is the volume, approximately, taken up by ourselves and the space immediately around us – it is roughly the volume occupied by a coffin, which is perhaps an appropriate symbolic unit when we are talking about the destruction of the planet. Once we have this image in our minds we can then start to relate that 'coffin's worth' of $CO_2$ to the exhaust gases of a 2 litre car travelling 10 miles, or to the emissions resulting from leaving on a 100 tungsten watt electric light bulb for a day (or a fluorescent bulb with similar light output for a week). We can look at a pound of strawberries from Israel and recognise that it costs us and the world that same coffin's worth of $CO_2$ to bring it to London.

We can also relate this to our current global 'earthshare' of manmade $CO_2$ emissions per person – being 4,000 of those coffins every year. In the UK each one of us is responsible for nearly 10,000 coffins and America is responsible for 20,000. In a sustainable future our emissions should be less than 2,000 coffins per year which, with an irony that was discussed at great length amongst the Cape Farewell crew, was roughly the amount of $CO_2$ that we had each expended on our return trip to Svalbard, over the course of just one week.

The only preoccupation I brought with me to Svalbard was to use this volume as part of a sculptural statement in snow and ice. Antony Gormley and myself both had an interest in constructing forms using simple blocks that we could cut from the snow, regularised and Euclidean – quarrying a material that had been there for months rather than millennia, and creating space and volume that made simple temporary statements focused around our individual and shared preoccupations.

We discovered that we could saw quite precise blocks with a density somewhere between lightweight concrete and polystyrene, but in our building techniques we had to be very precise because the snow itself, being very dry, did not lend itself to being used as 'mortar'.

Our discussions and reference points over the three day period ranged from the powerful primitive architectural forms of Egypt and Peru, Mycenae and Pylos, through to our experiences of the quarries at Bath and Carrara. We created a community of forms – a primitive block cut from the virgin snow, a vertical standing room of similar proportions again related to human dimensions, and a snow cave with a significant approach route and threshold, again based on orthogonal cuts into the organic drift of wind blown frozen snow. We found that we developed a strong relationship with the site, a longing to be out there digging and creating, whilst also absorbing the extraordinary scaleless white landscape that surrounded us. We were blessed with brilliant sunshine that provided intensely sharp and long shadows so that brought everything that we did into a higher resolution. We were delighted with the experience of what seemed like a 10° temperature difference between the inside and outside of the snow cave. It was essentially a sensory experience, working hard and playing hard to counteract the experience of being at -27°C, and producing work that was derived from individual preoccupations and joint collaboration and the inspiration of site and material.

The abstract body form enclosures had, for me, a further significance. Richard Feilden, my closest friend for 35 years (and partner for 27) was originally to have been a member of the Cape Farewell team, and I stood in for him only following his tragic accidental death over the New Year holiday. So the sarcophagus block, the first volume cut out of the snow, seemed to take on the character of an eloquent memorial to Richard. Intriguingly the whiter and lighter top layer of snow that was part of the natural formation gave it a natural 'lid'. When Antony and I collaborated on the vertical version of this volume, what emerged was a made place that was much more to do with light and life rather than death. Standing sentinel over the icebound fjord and bathed in sunlight, this enclosed void seemed even more of an appropriate place for Richard to inhabit. Our three Made Places – Block, Standing Room and Shelter are all reflections of the human form that represent a transient statement in what may turn out to be an all-too transient landscape.

⊕ **Sarcophagus** 2005 / 78°30N, 16°30E

'ONE KILOGRAM OF $CO_2$ AT ATMOSPHERIC PRESSURE OCCUPIES 0.54 OF A CUBIC METRE. THAT IS THE VOLUME – APPROXIMATELY, TAKEN UP BY OURSELVES AND THE SPACE IMMEDIATELY AROUND US – IT IS ROUGHLY THE VOLUME OCCUPIED BY A COFFIN, WHICH IS PERHAPS AN APPROPRIATE SYMBOLIC UNIT WHEN WE ARE TALKING ABOUT THE DESTRUCTION OF THE PLANET.'

⊕ **Ice Towers at the Ice Garden, Bodleian Library, Oxford** 2005 / 51°33N, 2°30W

**Decentralised Energy and Climate Change** Charlie Kronick, Chief Policy Adviser & Leader Climate Change Campaign, Greenpeace UK

Electricity production in the UK is responsible for nearly 40% of our carbon emissions. This is the UK's single greatest contribution to climate change. It need not be so. Our centralised model of production and transmission wastes an astonishing two-thirds of primary energy inputs, requiring us to burn far more fuel and emit far more carbon dioxide than necessary. It is hard to imagine a more wasteful and inefficient model than that which currently services the economies of the 'developed' world.

In our existing system, electricity is produced in a small number of large power stations, and then distributed to where it is needed. Because the power stations are generally far from centres of demand, the heat that is produced when fossil fuels are burnt is not used, but vented up chimneys or discharged into rivers. This heat loss alone represents a wastage of over 60% of the total energy contained in fossil fuels. Further losses occur as the electricity travels along the wires of the transmission and distribution systems. In total, the energy wasted at the power station and over the wires is equal to the entire water and space heating demands of all buildings in the UK – industrial, commercial, public and domestic. This is a nonsensical way to run our economy and power our lives.

But there is an alternative. In a decentralised energy system, electricity would be generated close to, or at, the point of use. When fossil fuels were burnt the heat would be captured and used. Buildings, instead of being passive consumers of energy, would become power stations, constituent parts of local energy networks. They would have solar photovoltaic panels, solar water heaters, micro wind turbines, heat pumps for extracting energy from the earth. They might also be linked to commercial or domestic combined heat and power systems.

An ideal energy system including decentralisation consists of these main elements:
- Heat and electricity generation close to the point of use allows the maximum benefit from any fuel used. Generating heat and power together increases the value of the fuel enormously. Currently around 2/3 of energy in the UK is lost as wasted heat at the power station or in long distance transmission.

- Renewable energy technologies like wind, wave, tidal and solar power offer carbon free energy and the lowest possible environmental impact. They use no fuel, relying only on endless indigenous resources, like wind and waves, in which the UK is rich.

- Increasing energy efficiency at its point of use in the home, in factories or in businesses is the cheapest and most effective way to cut carbon emissions and can reduce energy demand. Reducing demand is the most effective way to reduce fuel use and energy dependency. A decentralised energy system, which gives people more active ownership of their energy sources, is a crucial element in effectively stimulating efficiency in the use of that energy.

A vision of the future can be found in the town of Woking. By decentralising energy generation, capturing and using the waste heat from its power plants and improving energy efficiency, the council there has cut $CO_2$ emissions from its own buildings by an impressive 77% over the last 15 years. It is clear from this example that decentralising energy generation has the potential to cut our emissions drastically.

This radical transformation of our energy system sounds attractive but expensive. But in fact, decentralising our energy sources, instead of replacing our current centralised system, would actually save money in the long run. According to the International Energy Agency, over the next few decades the European Union will spend $500 billion on modernising and replacing transmission and distribution networks. The opportunity to avoid many of these costs means that decentralised energy makes economic, as well as environmental, sense.

The reasons that decentralised energy has such advantages over centralised generation are threefold:
- Generating electricity near the point of use reduces the electricity network required, so it avoids network losses and reduces the transmission and distribution costs of power plants. This is especially relevant to the UK, because most demand growth for electricity over the coming 20 years is expected in urban areas like the Southeast. In these areas the national grid is already close to capacity and so significant new investment to upgrade it would be required for new centralised generation.

- The fuel efficiency of decentralised energy is generally higher than of centralised generation, because localised energy generation allows for the use of both the heat and power outputs of the process. Consequently, a decentralised energy system requires less generating capacity and uses less fuel to meet the same electricity demand.

- Decentralised energy requires less backup capacity than centralised generation because, unlike a system consisting of a few large power plants, a system of many small generators cannot suffer a major impact from the outage of a single generator. This also means that electricity supplies under a decentralised system are more secure.

Decentralised energy also offers a way forward for developing nations and for the emerging economic giants like China and India. It is sometimes claimed, fatalistically, that efforts to stabilise the climate will be overwhelmed by China burning its coal reserves. But developing a decentralised energy system in response to its burgeoning demand for power would mean that China's emissions would be less than half those from a centralised system.

Decentralising energy offers a compelling alternative vision. To illustrate this potential, Greenpeace has used the World Alliance

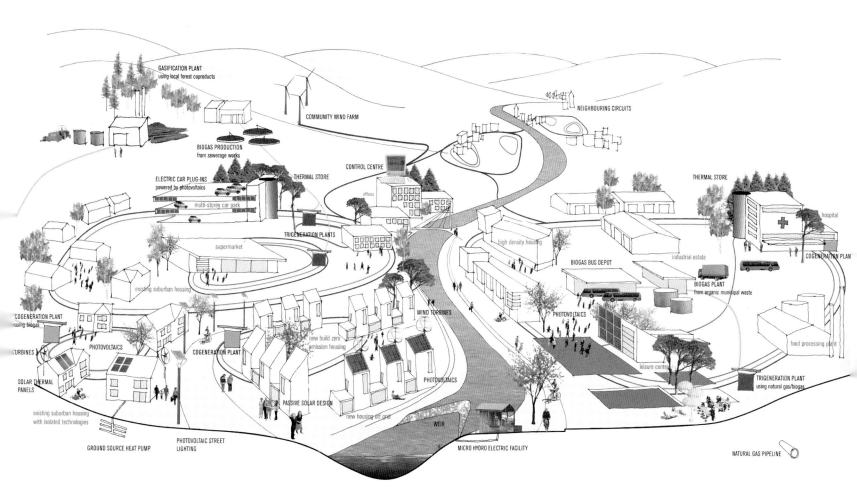

◆ **Model Village** Greenpeace

for Decentralised Energy (WADE) model, which compares traditional centralised energy systems to decentralised ones. The model has recently been used by the UK Foreign Office to project China's energy future, and the European Commission to investigate the options for the EU. It is currently being used by the German Environment Ministry to investigate the potential for decentralised energy there.

Greenpeace commissioned WADE to compare two basic scenarios. First, a 'business as usual' centralised scenario in which existing nuclear plants are replaced with new nuclear stations. Because investment has already gone into renewing the centralised transmission system, this scenario assumes that coal-fired power stations are also replaced upon retirement by new centralised plants, mainly gas-fired. Secondly, a decentralised scenario in which there is no nuclear new-build, and nuclear and coal stations are replaced predominantly by decentralised generation including gas- and biomass-fired combined heat and power (CHP) and localised renewables.

The model results show that the decentralised scenario is cleaner, cheaper and more secure. $CO_2$ emissions are 17% lower. The capital costs are lower by over £1billion in the UK alone, because the enormous cost of upgrading the transmission and distribution system has been significantly reduced, and the retail cost of electricity is lower. Gas use is lower by 14%, leading to lower dependency on imported fuel. This counter-intuitive result is explained by the fact that in the centralised scenario gas is burnt in inefficient power stations, and much of the total energy value in the fuel is lost in the form of waste heat going up the cooling towers, whereas in the decentralised scenario it is primarily burnt in more efficient CHP stations.

The claim that any meaningful response to climate change will cripple economies around the world and leave us with only 'difficult choices' if we wish both to reduce $CO_2$ emissions and enjoy security of supply, is simply untrue. Decentralised energy gives us a real choice now. What are we waiting for?

**⊕ Monolith** 2005/ 78°30N, 16°30E

'WHEN ANTONY AND I COLLABORATED ON THE VERTICAL VERSION OF THIS VOLUME, WHAT EMERGED WAS A MADE PLACE THAT WAS MUCH MORE TO DO WITH LIGHT AND LIFE RATHER THAN DEATH.'

**Journal 14/04/2005** Antony Gormley, Artist

What we have done in a tiny way is make a construction that conforms (or attempts to conform) to the absolutes of Euclidian geometry. In some sense, this talks about the human animal and the way that the human animal insists on making shelters according to abstract principles. No other animal does that. Here it is a foreign object, a space ship. Being in the snow cave, for me, is so powerful because of the relationship between the made human world and the inherited Earth – the Earth out there in that blue light that goes on forever. For me it has been a very precious reinforcement of something I feel deeply of how we are a gnat on a nose of a totally indifferent universe.

These three places are all made and do not seek to describe the body but indicate its place in an un-inscribed Arctic environment. Taken individually, the block indicates a relationship between the individual body and a planetary body-mass. The luminous void chamber is a vertical space that indicates consciousness and the shelter establishes the necessity of a collective body. Together, all three constitute a continuum of places that the human needs to dwell in: the physical space of the body, the imaginative space of consciousness, and the collective space of fellowship.

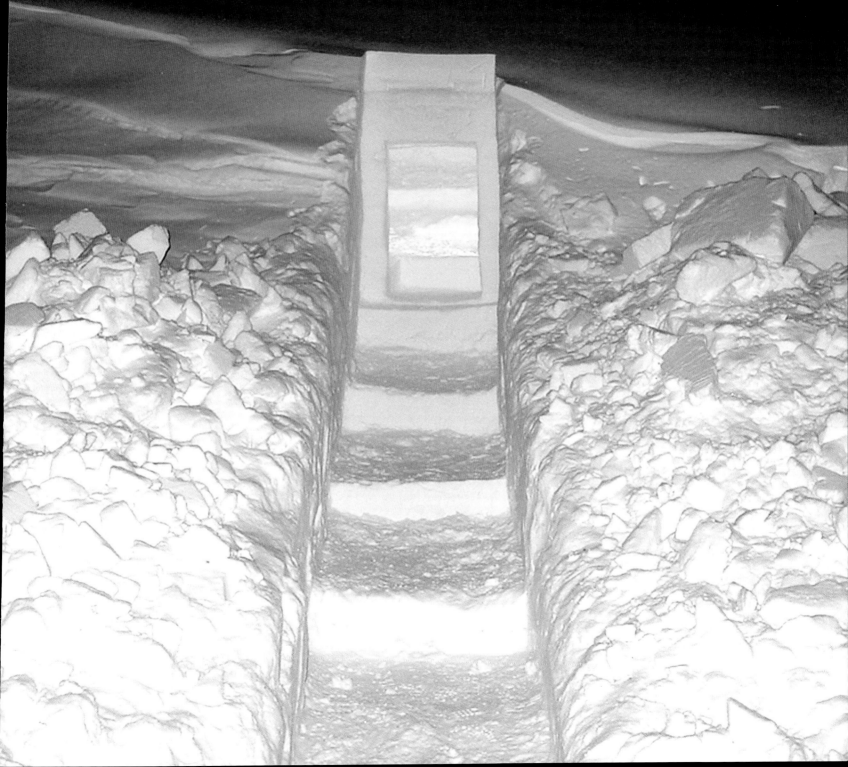

⊕ **Three Made Places** 2005/ 78°30N, 16°30E

**⊕ Messenger** David Buckland, Artist

BLACK OIL HAS BEEN FOUND UNDER
THE WHITE ICE OF THE HIGH ARCTIC.

WE SEEM UNABLE TO PROTECT THIS
WILDERNESS FROM EXPLOITATION,
UNABLE TO CEASE BURNING THE OIL
THAT DESTROYS IT.

THE SHADOWY FORM OF A PREGNANT
WOMAN WALKS OVER BLACK ICE, HER
BODY SERVING TO REMIND US OF OUR
RESPONSIBILITY TO UNBORN GENERATIONS.

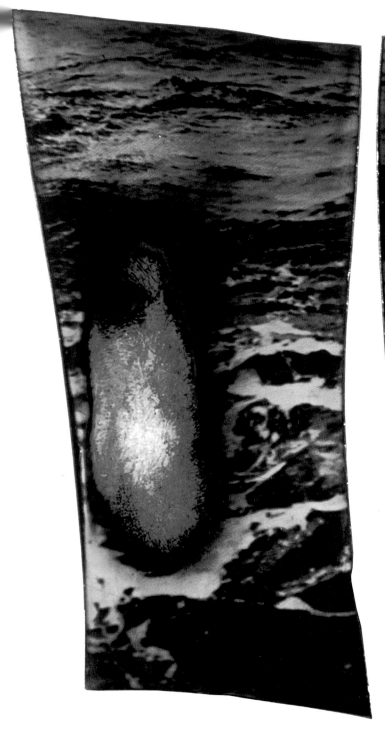

⊕ **Black Ice** Collodion on Black Glass, 2005 / 78°30N, 16°30E

**Journal May – June 2005** Gretel Ehrlich, Novelist, Poet & Anthropologist

**May 27**
Left Tromsø, Norway at a time when, in London it would be getting dark, but here it is light all night. When the mainsail is hoisted, the boat tilts to starboard, and the halyards slap and vibrate. Five sails go up. A hard westerly wind fills them. I'm on the Noorderlicht, a 150-foot gaff-rigged schooner built in 1910, traveling north on the Barents Sea to an Arctic archipelago between latitude 76 and 80 degrees North, called Spitsbergen. As soon as we leave the last bit of Norwegian coast, the seas become heavy and rolling. There are 20 of us on the Cape Farewell expedition, a mixed group of scientists, naturalists, writers, artists, photographers and two film crews. We're going there to better understand the richness and fragility of the Arctic ecosystem, see for ourselves these Arctic islands, if and how drastic climate change is affecting land and water, and how the Arctic can be a canary for the global ecosystem.

Light all afternoon, all night. The seas grow heavier. We pass the last Norwegian lighthouse on a point of rock, then plunge north into the wild Barents Sea. House-sized swells lift up. The boat's name, 'Noorderlicht', is Dutch for Northern Lights, but I see none here. Instead we have grey skies, persistent rain, and an ocean that is bigger than we are. Because we'll be sailing day and night for two days, night watches are assigned by the Captain, Ted, and his first mate Maaike. Mine is from midnight to three.

When I come on deck, the seas are so rough, and the decks so slippery, I'm given a harness, and snap into the rail near the big steering wheel. Rain turns to sleet, then rain again. We can see because it's light, but we can barely stand. How easy it would be to go over the edge. Who would see?

Strong wind puts the rail near the water's edge. Below, the misery of seasickness begins to take hold. Despite persistent rain, I'm grateful for the fresh air. Midway through the watch Maaike brings tea and biscuits. Best to keep something in your stomach she says.

Almost everyone is sick. I cling to my harness, a bit like a monkey. Anything not battened down falls and breaks. My equilibrium unravels. Then I'm sick too. Oh misery … The first night merges into the second day. We are still under way, still keeling over in an 18 knot wind – 'a natural nine', Ward, the second mate calls it: 9 knots under sail. The Noorderlicht groans, creaks and shivers.

**May 28**
Too sick to write. Watery buildings are falling on me. The floor jumps. A bench slides. Glasses break. I'm shaking with cold. What day is it? I've slept sitting up in the upstairs cabin, but go on deck when I can. The bow bangs down into abysmal troughs then points straight up. Will we ever stop, will we ever get anywhere?

**May 29**
Three days have passed. I'm not sure if it's afternoon or morning. 'It's near midnight', Maaike says. She's at the helm. When I ask her how she started sailing she says, 'I was born on a house boat and it went from there.' She points to a white mirage straight ahead: it's land: 'Bear Island'.

The chart says that we're between latitude 74 and 75 degrees North. A line of Brannich's guillemots flies. They form a black line above the gold horizon. Everything here is made of lines, all dismembered and askew, picking up glints and hues, odd ducks, and round the clock light as they travel. It's midnight and we can see the sun for the first time. Here it hangs halfway between the top of the sky and the horizon.

Slowly an island comes into being. It is rimmed with ice, swathed by clouds, and ends in purple cliffs where sea birds nest, and fledge. The west side is snow-blasted. Sunlight breaks on cliffs. There are paths of green. A ring of ice encircles the rest of the land.

1.00 a.m. We move in closer. The ice pack is jammed up against the flank of the island. It is its own symphony of sloshing and sluicing. Swells rise and fall and the whole mat of ice and our boat rises and falls.

Sun makes a white string out of ice at the horizon. A pod of harp seals surfaces, bobbing up and down. Ice is the dynamic that makes everything move; or is it the sea, or the wind, or the tides, or the sun that we can now see?

'Bear!' Maaike shouts. A polar bear is running across the ice. He leaps from flow to flow, swims the crevasses, leaps again. Near the ice edge he jumps down on the beach and swats at an eider duck between two pieces of ice – misses – looks all over for it … gives up … strides away. His movements seem effortless, as elastic as ice. He jumps up on land and looks out. Nose up, he catches our multiple scents. 'What are we?' he must wonder – a red boat with people hanging over the rail. Now he stands. The human perfume is too delicious after a long winter. The bear jumps into the water and swims towards us.

Are we dinner or trouble? Because bears aren't hunted here, he's quite fearless. He's cautious but swims closer. Now behind the stern, he treads water, looks from side to side, then realises we are part of a machine, something metal … and gives up, swims to shore.

3.00 a.m. of whatever day this is. Falling water stacks up as ice. A seagull twirls. The ocean turns 360 degrees in a glass. We drift far to the west threading our way through ice. Where sea water has eaten the edge of the island, huge caves arch up. Above, one waterfall is needle-thin and disappears half way down the mountain.

Little auks sit side by side on bits of brash ice. I sleep for two hours, then get up again. Now we're going north and the sun is out. The two oceanographers, Val and Sarah, take measurements of warmth and salinity. It's here that the warm Gulf Stream slides up the west coast of Bear Island, while the cold North Atlantic current comes down the east side. When the two meet, the denser, saltier warm water buckles under the cold and turns: circulation begins. If the warming trend in our climate continues, and the Greenland ice cap melts, adding huge amounts of cold, fresh water to the oceans, the warm, tropical water will sink, Europe and the UK will freeze. A new ice age would begin.

Hot or cold? Which way are we going? Both directions at once? Hot and hotter? Ice and snowpack send heat from the sun back into space, thus keeping things cool. If all the ice and snow melt, the earth will become like a sponge, absorbing heat. Our water sources will quickly dry up, plants, humans and animals will die … Is this what is in our future? We can look at the history of the weather, but what's new is the human-caused pollutants as they conflate with the current warming trend.

Who and what is at risk? What will become of the seals, walrus, birds, fish, and polar bears? How is the polar North a canary for the health of the entire planet? How are we moved by what we see on this journey? How can the educational curriculum better prepare our young people to understand and act to maintain the health of the planet, how this place moves us, what we find beautiful in a place too many glibly describe as 'a desolate waste.'

Under way again. Rolling seas instead of ones that rock. Rock and roll must have been a sailor's term. Oceanic music. I love my middle of the night watch. The sun is in the north and casts an eerie light across the sea. The water is aquamarine – a thick blue-green broth. The night-sun blazes a white path from horizon to me.

**May 30**
Following the west coast of the most southerly island, I see mist-dripping peaks, low-slung tidewater glaciers unleavening themselves into whitecaps. In spring ice eaten by water equals the past in the future; but water eaten by ice represents past and future in the present. Glaciers are archives. They store time, pollen, bones, weather events, bodies. Ice is always teaching water about cold, and water is swallowing itself. It takes 1000 years for a drop of water to go through the ocean's global circulation. 'One drop goes a long way,' somebody says.

We search for safe haven – a place where the anchor will touch ground. Snow falls. Newly arrived, two eider ducks swim close together like newlyweds. Ted finds a place to anchor for the night. Calm water is the last thing I'll ever want. A pod of beluga whales swim fast away from us. Max, the sound artist, drops a hydrophone down and we listen to beluga song. Where they are and how many of them are singing we can't know. Their sounds carry a long way under water. But it is loud and clear: high notes in a rapid descent, followed by another and another. What are they saying?

**June 1**
Tides of bearded seal and beluga song come and go. Ocean currents clash. An iceberg is upended, sending waves on shore. A snowflake falls, moving the mass balance of a glacier toward accumulation, though later, it is lost as ablation takes over. Every moment of life is a delicate balance. We are so ignorant of the fragility of earth, ice, air, fish, water, ourselves. Jason Roberts, an Australian adventure in Spitsbergen, says, 'The land has been beautifully preserved here. But they forgot to include the ocean. Without a healthy sea, the Arctic ecosystem collapses immediately. Bears, walrus, bird, seals need fish. Fish need healthy water. They need to be left alone. The whole world is overfished now. The food chain in the Arctic is very short – only three steps. Take one element away and everything goes. How could people have been so short-sighted?'

We drift through icebergs. Seal song soaks up through the hull. By the time we finish this two-week voyage we will have traveled 1,020 miles. Now we have our sea legs under us. We debate the variables having to do with global warming. How much is natural cycle? How much provoked by human-caused pollutants from industrial waste, deforestation, overgrazing, burning? Is sea level rising? Is the warm Gulf Stream going under? Of the 160,000 glaciers in the world, most are retreating. Ted has marked how much the glaciers in Spitsbergen have melted back since he began sailing here. Directly across the North Atlantic, the Greenland icecap is pouring meltwater into the sea. As the salinity of the ocean decreases, the denser, saltier water from the tropics – the Gulf Stream – that warms Europe, begins to sink, and might come to a stop altogether. If this happens, the UK and Europe will turn into an ice box. We are looking at a landscape that is vanishing …

**Ice Events** Max Eastley, Artist

'ON THE BEACH, I WAS RECORDING THE ICE. IT GIVES OFF THESE CRACKLING NOISES, WHICH ARE THE SOUNDS OF BUBBLES THAT HAVE BEEN COMPRESSED BY THE GLACIER. WHAT YOU ARE HEARING IS A 10,000 YEAR COMPRESSED EVENT BEING RELEASED FOR JUST A TINY INSTANT.'

◆ **Wind Music** 2004 / 78°55N, 12°00E

'YOU THINK IN TERMS OF NATURE AS BEING SLOW.
BUT YOU'RE MISSING IT ALL THE TIME. IT'S VERY, VERY FAST.'

'I COULDN'T SAY THAT ART WILL SOLVE EVERYTHING; IT WON'T SOLVE ANYTHING. BUT THERE HAS TO BE A KIND OF AMALGAMATION OF EVERYBODY ALL LOOKING IN THE SAME DIRECTION. FIRST, YOU HAVE TO CONVINCE PEOPLE THAT CLIMATE CHANGE IS HAPPENING, SECOND, WHAT WE CAN DO ABOUT IT AND, FINALLY, HOW LONG HAVE WE GOT?'

Max Eastley 57

'IT HAD A SWORD OF DAMOCLES QUALITY. YOU COULD SEE THE DRIPPING WATER, BUT YOU COULDN'T SEE WHEN A STONE FELL, SO THERE WAS SUDDENLY THIS BANG. IT WAS REALLY SHOCKING. ONE OF THEM MADE A REALLY LOUD BANG, AND THERE WAS A CRACK. THAT'S JUST WHAT ICE DOES. WHEN YOU'RE NEAR THOSE GLACIERS, YOU'LL HEAR A SOUND LIKE A FIELD GUN GOING OFF'

⊕ **Glacier Ice at the Ice Garden, Bodleian Library, Oxford** 2005 / 51°33N, 2°30W

**Journal 12/03/2005** Gautier Deblonde, Artist

Today three of us went to Pyramiden, a Russian mining town closed since 1998. You don't really notice it at first but there is no activity. Its buildings still stand, deserted, it feels like a film set. You come across the swimming pool, the sports club, the cinemas, and everything has been left behind. In the library all the books are on the shelves. In the indoor basketball court the balls have been left on the floor. We go to the cinema, to the projection room, and all the films are still stacked there. It feels as if you can hear the people who lived there – just like when you're in a room where a party has just finished, and there's a lingering sense of the voices and the music. There is something magical about this place. Everything seems as if it is waiting to be used. It's all on the verge of happening, you would just have to put the heater on and life would start again.

I'm going to Barentsburg tomorrow, established in 1932; it's the last Russian settlement in the Arctic. I've always wanted to stay there so I've booked a room in the only hotel for a week. It proves to be a small town, a few streets and a mine, which is grim, run down and melancholic. The snow is covered with coal dust and you hardly see anyone outside. Six hundred people from Russia and the Ukraine work there on two-year contracts: 400 men and 200 women all come for the money. Many families are divided; they leave their children with relatives in Russia while they work at Barenstburg. There is nowhere else to go while their contract lasts; they are trapped in this town surrounding by hard, fragile icy mountains.

The work and the quality of life has been the same for years, they live under almost impossible circumstances but they are, as a community, incredibly generous with their time and spirit.

When I talk to them about global warming, they say that it does get warmer sometimes, that it rained last summer for the first time, the glacier has shrunk, but they have to work, life goes on.

Gautier Deblonde

**Barentsburg** 2005 / 78°05N, 14°00E

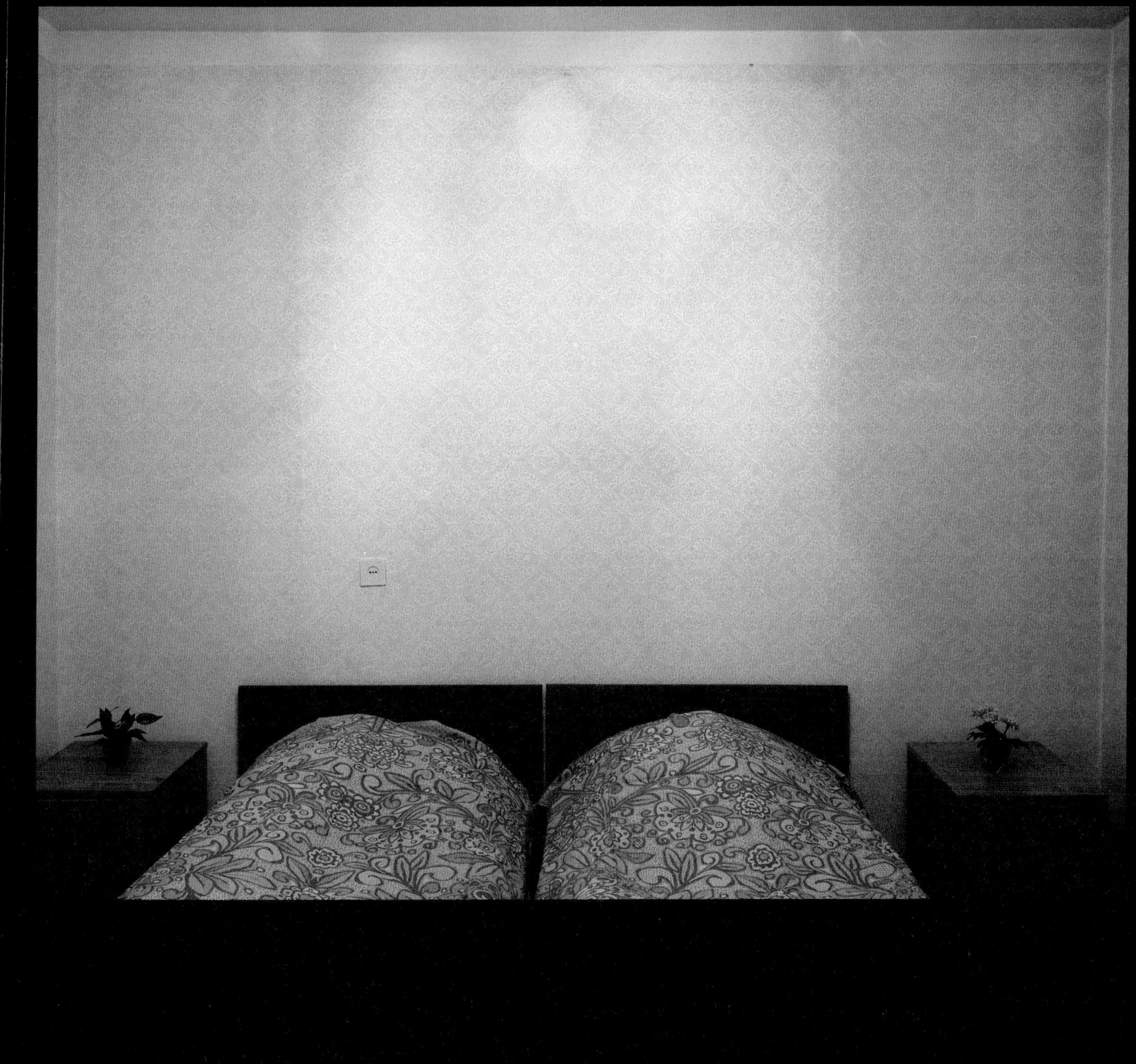

72

**The North Atlantic Carbon Conveyor** Dr. Valborg Byfield, Oceanographer, National Oceanography Centre, Southampton

The northern end of the Atlantic conveyor is also important for the ocean's ability to take up and store carbon dioxide – the foremost gas in an array of greenhouse gases – so called because they trap heat in much the same way as a glasshouse, by letting sunlight in to warm the Earth, but stopping heat radiated by the Earth's surface from escaping to space. The natural greenhouse effect makes the Earth some 30 degrees warmer than it would otherwise be. Humanity is adding to this effect; the concentration of carbon dioxide in the atmosphere is now higher than at any time in the past 650,000 years.

Each year we release about 8 billion tons of carbon into the atmosphere, 6.5 billion tons from fossil fuels and 1.5 billion from deforestation. Yet less than half that – 3.2 billion tons – remains in the atmosphere to warm the planet. Where does the rest go? It is a mystery that science has not yet fully solved. We know that the ocean, forests, peat bogs, and grasslands must be acting as carbon sinks, helping to slow the build-up of carbon dioxide in the atmosphere, and delaying its effect on climate. We estimate that about half, perhaps more, of the missing carbon ends up in the ocean.

Ocean water holds 50 times as much carbon dioxide as the atmosphere – the vast majority of this (over 97%) in the deep ocean. Carbon dioxide stored in the deep ocean may take hundreds of years to return to the surface, effectively buying us more time. And this is where the Atlantic conveyor comes in.

Colder, fresher water can hold far more of this greenhouse gas than warmer, saltier water, so, as the Gulf Stream water cools on its way north, it also absorbs carbon dioxide from the atmosphere. The colder, fresher Arctic water is also rich in dissolved gases. When these waters mix and sink, they transport carbon dioxide into the deep ocean, where it will remain away from the atmosphere for hundreds of years. Thus a key route for atmospheric carbon dioxide into the deep ocean is found at the edge of the Arctic, at the northern end of the Atlantic conveyor.

▶ **The greenhouse effect**

**1** Energy from the sun arrives at the top of the atmosphere mainly as short-wave visible radiation.
**2** This visible light passes easily through the atmosphere to warm the Earth's surface.
**3** Some is reflected back into space by clouds, snow and ice, which may reflect from 50-90% of incident light, depending on how fresh it is. Open ocean and snow-free land absorbs much of this radiation; oceans reflect from 4-5% to 50-60% depending on sea state and sun angle (at low sun angles more light is reflected).
**4** The solar energy input is balanced by long-wave (infrared) radiation from the Earth's surface. This radiation is absorbed by greenhouse gases (carbon dioxide, methane, water vapour) and by clouds, heating the atmosphere. The amount of infrared radiation emitted by the Earth's surface depends on its temperature; the warmer the surface, the more energy it radiates. As the concentration of greenhouse gases increases, more long-wave radiation is trapped in the atmosphere and converted to heat, so the Earth warms up. However, as the Earth warms, the energy it radiates increases, until the loss again balances the incoming energy from sunlight. Thus the average temperature of the Earth's surface is controlled by two factors: the amount of energy we receive from the sun, and the concentration of greenhouse gases in our atmosphere.

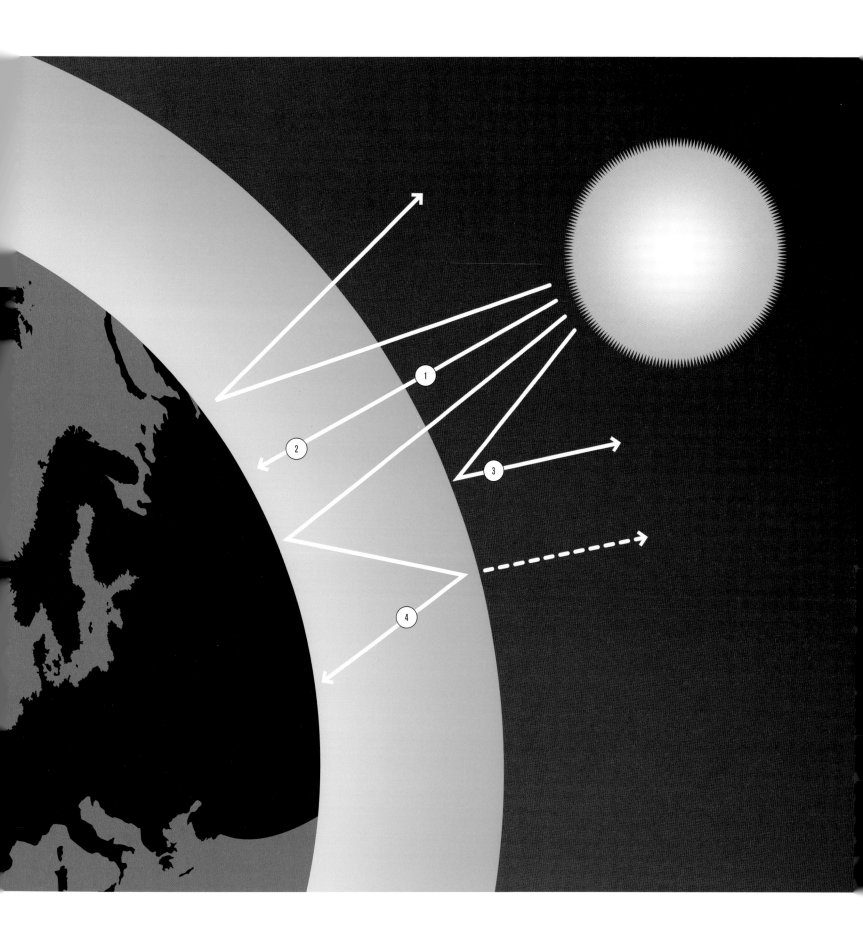

**Marine Snow** 2004 / 78°30N, 11°56E

### Marine Snow

So-called because it sinks slowly through the water, much as snow falls through the air. The whitish 'snow-flakes' can be a few millimetres across, and consist of dead and decaying phytoplankton, with zooplankton faecal matter and remains, all colonised by bacteria. They sink at rates from a few tens of metres per day to several hundred metres per day in contrast to phytoplankton cells which individually sink at no more than 0.1 to 1 metre per day. Marine snow starts to arrive at the sea floor soon after a plankton bloom, and is the basic food for a large number of bottom-dwelling animals. NOC/R.Lampitt

### The biological carbon pump

Another key route for carbon to enter the deep ocean is known as the 'biological carbon pump'. Microscopic marine plants, the phytoplankton live in the sunlit world near the sea surface. Like all plants, they absorb carbon dioxide as they photosynthesise, and store carbon away as they grow. Much of this carbon is released back into surface water after a few days or weeks, as the plankton die and decompose, but some sinks into the mud on the continental shelves or into the deep ocean.

This rain of debris, known as marine snow, feeds an enormous web of deep-sea and bottom-dwelling animals. Nearly all the carbon that reaches the deep sea in this way is recycled back into carbon dioxide, and will eventually return to the surface and the atmosphere. But some is buried in sediments and returned to the geological cycle.

The biological pump is active in all plankton-rich, productive areas of the oceans, and the regions where cold, nutrient-rich polar water meets the warmer nutrient-poor subtropical water are among the most productive in the world.

**The Ocean Carbon Cycle**

**1** The surface ocean exchanges $CO_2$ with the atmosphere.

**2** Plants in the ocean (mainly microscopic phytoplankton) take up $CO_2$ and water, turning this to organic carbon in cell tissues.

**3** This carbon is returned to the water as $CO_2$ in respiration – when plants and animals break down food molecules for energy and emit $CO_2$ gas and other byproducts. The organic carbon produced by marine plants can take several paths before re-entering the surface ocean. When plankton and animals die they are broken down by micro-organisms that feed on dead organic matter, and release carbon as $CO_2$. In the surface ocean carbon is continuously cycled in this way over periods ranging from a day to a few years.

**4** Some of the organic carbon sinks into the deep ocean as plants and animals die.

**5** Decomposition in deep water and bottom sediments release $CO_2$, which may take hundreds of years to return to the surface, effectively creating a second carbon cycle that turns over periods of hundreds to a thousand years.

**6** Some of the carbon is buried so deep in bottom mud that it is not released, but returns to the geological carbon cycle.

**7** It may take millions of years to return to the atmosphere through geological events such as volcanic eruptions. Humanity's burning of fossil fuels remove carbon from the geological cycle and releases it as $CO_2$ to the fast-spinning cycle between surface ocean and atmosphere. This happens at a much faster rate than the Earth system can return it to the geological cycle, and the as a result there is a build-up of this greenhouse gas in the atmosphere.

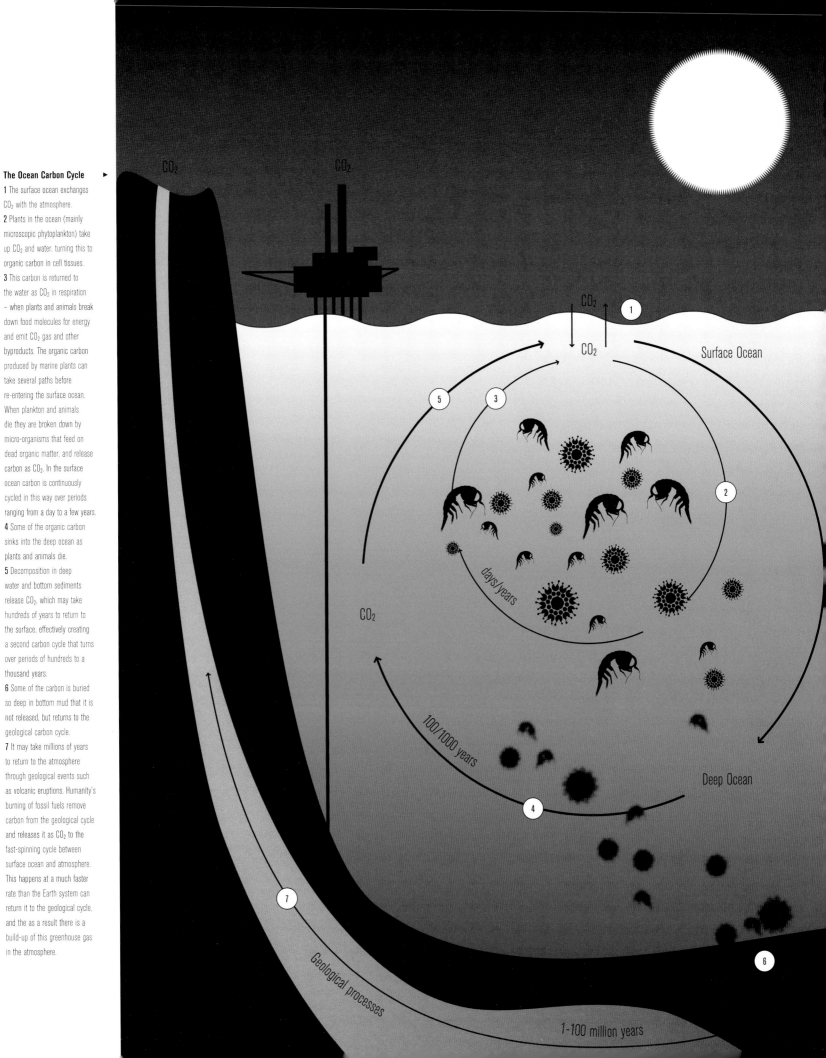

⊕ **Moffen Island** 2004 / 80°05N, 14°58E

▶ **Arctic Foodweb**

In this simplified food web most of the animals shown represent broader groups. The most important plants are the phytoplankton, but locally ice algae can also be important. The main herbivores are the zooplankton, including larvae of many sedentary bottom dwellers; some molluscs also filter algae from the water. Primary carnivores include larger invertebrates, mainly crustaceans (shrimps, krill, crabs); in their turn these are eaten by larger carnivores – fish, birds, and mammals. At the top of the food-chain are the top carnivores: polar bears, toothed whales and many of the gulls and birds-of-prey. Animals also feed on the droppings (faecal pellets) of other animals, and on the remains of dead organisms (known as detritus). For some, particularly smaller animals on the sea floor this is their only source of food. Some examples: zooplankton eat small particles (dead and alive) by filtering them from the water, benthic invertebrates (molluscs, crabs, various worms) filter particles from the water, or pick them out of the sea bottom mud; the Arctic fox eats carrion from seal kills left by polar bears.

### Life at the Arctic Front

As the first Cape Farewell expedition crossed the Devil's Dance Floor on its way to Svalbard, Arctic blooming was about to begin. Flying north alongside the Noorderlicht were flocks of petrels, sheltering from the gale by flying between the waves, getting lift from the wave crests. For over two days these birds seemed the only life in a vast world of grey, doggedly heading north and never stopping to feed or rest. Then, on the third day, everything changed. Instead of petrels skimming the waves, there were thousands of gulls and other birds bobbing gently around us. Soon the shout went up 'whale!' We soon realised the sea was teeming with life – fish, seals, dolphins, humpback whales. This was the Arctic front – that swirl of eddies where polar and Gulf Stream waters meet and mix – one of the most fertile areas in the open ocean.

The Arctic winter is one long, polar night, so Arctic plants produce most of their biomass during the long days of summer. This is also when the animals that depend on them for food do most of their feeding. The diet of Arctic animals is varied, particularly among mammals and birds. The real foodweb is more complex than shown. Higher level predators may catch animals from more than one level below them in the food chain: baleen whales often chase fish, or engulf them when filtering the water for krill; gulls eat mainly fish, but also the odd egg or chick from other fish-eating bird species if the opportunity arises. Fish and crabs start life as larvae among the zooplankton, and may at that stage be food for other animal species, which they may go on to prey on as they grow. Even phytoplankton may sometimes turn herbivore.

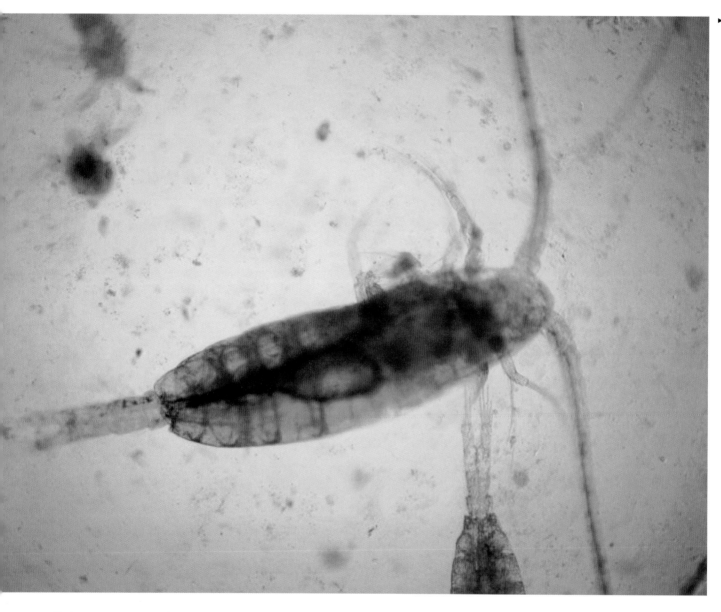

▶ **Phytoplankton Bloom**
This true-colour satellite image, taken by the MODIS instrument onboard NASA's Terra satellite on August 13, 2002, is typical of the Arctic summer. Phytoplankton blooming in the Barents Sea at the edge of the Arctic Front, are clearly visible as swirls of turquoise water against the deeper blue of less plankton-rich areas of the ocean. To the north Svalbard is mostly cloud free, but still partly covered in snow. Bear Island, halfway between Svalbard and North Norway may also be glimpsed through a gap in the clouds.

Credit: Jacques Descloitres, MODIS Rapid Response Team, NASA/GSFC

⊕ **Zooplankton** 2004 / 78°30N, 11°56E

## When the Arctic Blooms

Phytoplankton are the grass of the sea. Along with ice algae, they are the basis for all other life, from microscopic zooplankton to giant whales, from fish and crabs to birds, seals and bears. Even life on land depends to some extent on the productive ocean for nourishment. In the short Arctic summer, vast blooms of phytoplankton can spring up over a few days to become so bright and extensive that they are clearly visible from space.

As sunlight returns, and the ice begins to melt, zooplankton and fish spawn, birds return to nest, seals and walrus give birth, bears emerge with their young from under the snow – all ready to take advantage of the short, sunlit season when phytoplankton and ice algae bloom.

## From Gaia to Geopolitics
Dr. Tom Wakeford, Ecologist and action researcher

Today you will have almost certainly inhaled an atom of carbon exhaled by Julius Caesar, when he uttered the question 'Et tu Brute?' to his treacherous aide. Now multiply your breathing by the respiration of every plant, fungus, bacteria, human being and other animals. You do not need a calculator to conclude that organisms have, by their very existence, exerted a powerful influence over the global climate. While walking in the snow today it struck me that it is exactly forty years since the British chemist and inventor James Lovelock published a paper in the journal 'Nature' that made this fundamental insight into a theory of global interconnectedness and a keystone of our current understanding of climate change.

Lovelock had been working for NASA on methods by which life could be detected on other planets. Long before the first landing on our moon, most space scientists had discounted the possibility that life existed there. They were more hopeful about Mars. Lovelock's ability to transcend the ordinary boundaries of scientific disciplines has allowed him to provide science with some extraordinary insights. His autobiography records how his trans-disciplinary approach sometimes led him to make mistakes, particularly when, in the 1950s, he recommended that all of Britain's hedgerows should be grubbed up to make agriculture more efficient. Many still believe that his unquestioning advocacy of nuclear power as a solution to climate change is based more on faith in technology than sober analysis.

Yet he was prophetic when, in 1962, Lovelock decided that there was no life on Mars. The data he had gathered to answer this question had also started him thinking about an ability life might have to collectively regulate the atmosphere over a period of millions of years. The planets either side of Earth, Mars and Venus, have virtually no oxygen in their atmosphere. Earth, the only planet with life, has maintained a fifth of its atmosphere as this highly reactive gas for hundreds of millions of years. The Earth, reasoned Lovelock, must have regulated its atmosphere – just as the human body regulates the concentration of oxygen in its blood supply – via a process called homeostasis. Even after he had, together with his collaborator Lynn Margulis, developed his idea into a rigorously argued hypothesis, many of their fellow scientists dismissed the idea. Whatever they thought of the science, many were put off the theory's name. In a marriage of science and art reminiscent of Cape Farewell, Lovelock took the advice of the novelist William Golding and named his idea after the ancient Greek goddess of the Earth, Gaia.

The late 20th-century flowering of Gaian thinking among scientists draws on humanity's eternal theorizing about the living Earth. The Gaia Hypothesis revived ideas of reverence for the non-human world that prompted its vilification by ultra-rationalists, such as Richard Dawkins. The key tenets of the hypothesis have latterly been rehabilitated by evolutionary theorists, such as E.O. Wilson, W.D. Hamilton and John Maynard Smith. Many of Lovelock's insights mirrored those made during the 1920s by the Russian polymath Vladimir Vernadsky. Both treated the Earth holistically, without negating the principles of rigorous research, yet were marginalised by their contemporaries in the world of science: Vernadsky because his ideas were expressed in Russia under the Soviets, Lovelock because he and his co-author of the hypothesis, Lynn Margulis, built ecologically grounded versions of Darwinism against a prevailing scientific culture of genetic reductionism.

Working with his then student Andrew Watson, Lovelock explained the mechanism behind his theory with his model, known as Daisyworld. Using a simple computer programme, it described an imaginary world that was only made up of two kinds of daisy. It demonstrated how homeostasis on a planetary scale could arise by pure Darwinian natural selection. The late Bill Hamilton, an evolutionary biologist, later proposed a mechanism whereby the regulatory impact of life on the Earth system could occur not just inside a computer, but in nature.

Hamilton was initially struck that microbes found in the tropical ocean contained anti-freeze. Microbes are masters of eliminating unnecessary components of their metabolism. In seas where freezing is very rare, synthesising anti-freeze is a waste of valuable energy. The more he found out about the marine system from his chemist collaborator, Tim Lenton, the more he began to suspect that microbes were unconsciously orchestrating the elements for their own perpetuation. He proposed that the survival of these microbes were linked to their ability to control the climate. These specks of life were, he suggested, using clouds, wind and rain to carry themselves around the planet, like a global taxi. The anti-freeze was needed not in the sea, but for the sub-zero temperatures the microbes had to survive at altitude. If he was right, this would be the most biologically credible mechanism for Gaia ever discovered.

Twenty years ago, a group of scientists led by Lovelock had proposed that marine microbes were part of a global regulatory system that kept the climate stable. Most of these microbes produce a gas called dimethyl sulphide (DMS), which reacts with oxygen in the air above the sea and forms tiny solid particles. These particles then form a surface on which water vapour could condense to form clouds. Clouds keep the planet cool by reflecting solar radiation back into space. Lovelock argued that this process could create a self-regulating global thermostat. Warmer weather would increase microbial photosynthesis and therefore DMS output, seeding more clouds, which would block out the sun. Then, as the climate cooled, microbial activity and DMS levels would decrease. But to be evolutionarily as well as meteorologically stable, DMS production must enhance the survival prospects of individual microbes.

Hamilton's ground-breaking idea was that the microbes were releasing DMS, not as an act of selflessness for the good of the climate, but to get themselves into the air. He had already devised a computer model suggesting that dispersal was a high priority for all organisms, their third objective after survival and reproduction. He also knew that microbes, like fungal spores, had been recorded as making intercontinental journeys at heights of up to 50 kilometres. He suggested that, as DMS causes water to condense around the sulphate particle, it releases energy in the form of heat. This warms the surrounding air, which then starts to rise, taking the microbes with it. Hamilton and Lenton think that a simultaneously-produced chemical DMSP, would act as an antifreeze, stopping the cells dying in the upper atmosphere. Once there, the tiny sulphate particles trigger cloud condensation, which eventually causes rain, allowing the microbes' final fall back to Earth.

Biologists such as Hamilton have become part of a growing number of scientists that have strayed out of their original discipline to provide valuable insights to climate studies. In an era where climate degradation seems an increasingly dangerous threat, understanding the biological dimensions of our climate is an urgent issue.

Hamilton's theory also has implications for more than just marine microbes. Dispersal is just as important for organisms on land as it is in the sea. Among those species both small enough to become airborne are microbes living on leaf surfaces and the lichens that have so impressed the Cape Farewell artists on our voyage. Though much more research is needed, preliminary findings seem to suggest that both marine and terrestrial microbes of this sort are found high in the atmosphere. Hamilton even proposed that they may perform their meteorological magic by working in teams.

As the triumph of bacteria over antibiotics has shown, many of the inventions meant to be magic bullet solutions to wipe out a disease have turned out to be more like boomerangs – they've come back to haunt us. Others may become so over time, as microbes catch up with those who would tame them. Gaia's message is similar. We are all ecologically bound up into a larger biological whole. This living system may or may not have the capability to regulate the parameters of the Earth that are crucial for life to continue. Even if the regulation occurs, humanity is neither in control of this regulation or necessary for its continuation. We continue to destroy our soils with industrial agriculture and burn up the atmosphere with our cars. If we buck this living global system too much, it could move to a state which is no longer hospitable to survival, first for our more vulnerable communities, and eventually all of us.

Even if we take seriously the need for humanity to take action to address our global climate crisis, the quasi-religious belief in the selfish gene by our political class frustrates many of our efforts. Richard Dawkins justifies unfettered free-market capitalism as if it naturally follows from the laws of nature. With remarkably little opposition, these ultra-Darwinists have helped create a twenty-first century economic system that values the pursuit of individual gain so highly that its social and environmental consequences become sidelined. Yet, as I charted in a book called Liaisons of Life, the same scientific evidence has been used by some of the world's most eminent scientific thinkers to build a far more holistic view of life – often dubbed Ecological Darwinism.

Encompassing Gaia theory, this worldview sees life as being as much based on the symbiosis of organisms rather than their mere competition. The lichens we saw thriving in such extreme conditions could only survive the because of the mutual dependence that had evolved between a fungus and an alga. Similarly, we can only survive the coming extremes of our global climate if we learn to live more symbiotically with the living things around us.

The society we create does not ultimately depend on the facts we collect, but on the values we hold. If we wish to become mutualistic nurturers of our environment rather than mere occupiers of it, we must embrace and understand our dependence on it. And if we are to successfully address the ecological genocide that many of the world's most vulnerable communities face, scientific analysis must be guided by the very un-selfish concept of universal human rights.

Some among a new generation of scientists give me hope about the future. They are passionate about human rights, modest about the certainty of their knowledge and include broader and holistic perspectives in their judgments than those that emerge out of traditional reductionism. For me, Cape Farewell is about bringing art and this new relationship to natural world together.

**Journal 11/04/2005** Siobhan Davies, Choreographer

It is incredibly hard to think what kind of movement would have any relevance up here. Any of the full range of movements that I am able to articulate in a warm studio seems grossly over the top in a landscape such as this. The best movement is to walk or to run. Walking gets us somewhere and keeps us warm. It is a simple action, but there is an efficiency about it and in many layers of clothing it became denser than we had ever experienced. Dance here is then reduced to the familiarity, the straightforwardness, the elegance of the walk and the run.

On Thursday morning most of us collected together to pace out the landscape for the film crew. It was a few moments of walking the landscape, exploring the rhythm, sound and changing size of the human body, before the company dispersed and went back to their own projects. If I think about the list of names that formed this short-lived dance company it makes me laugh with pride.

Some of us carried on with the walking theme by traversing an icy mountain, snow blowing along the ground, making it unstable as if floating. Our walking turned into slipping on ice or pushing our heels down into snow to help us keep upright while going steeply down hill. Every kind of putting one foot in front of the other had its use.

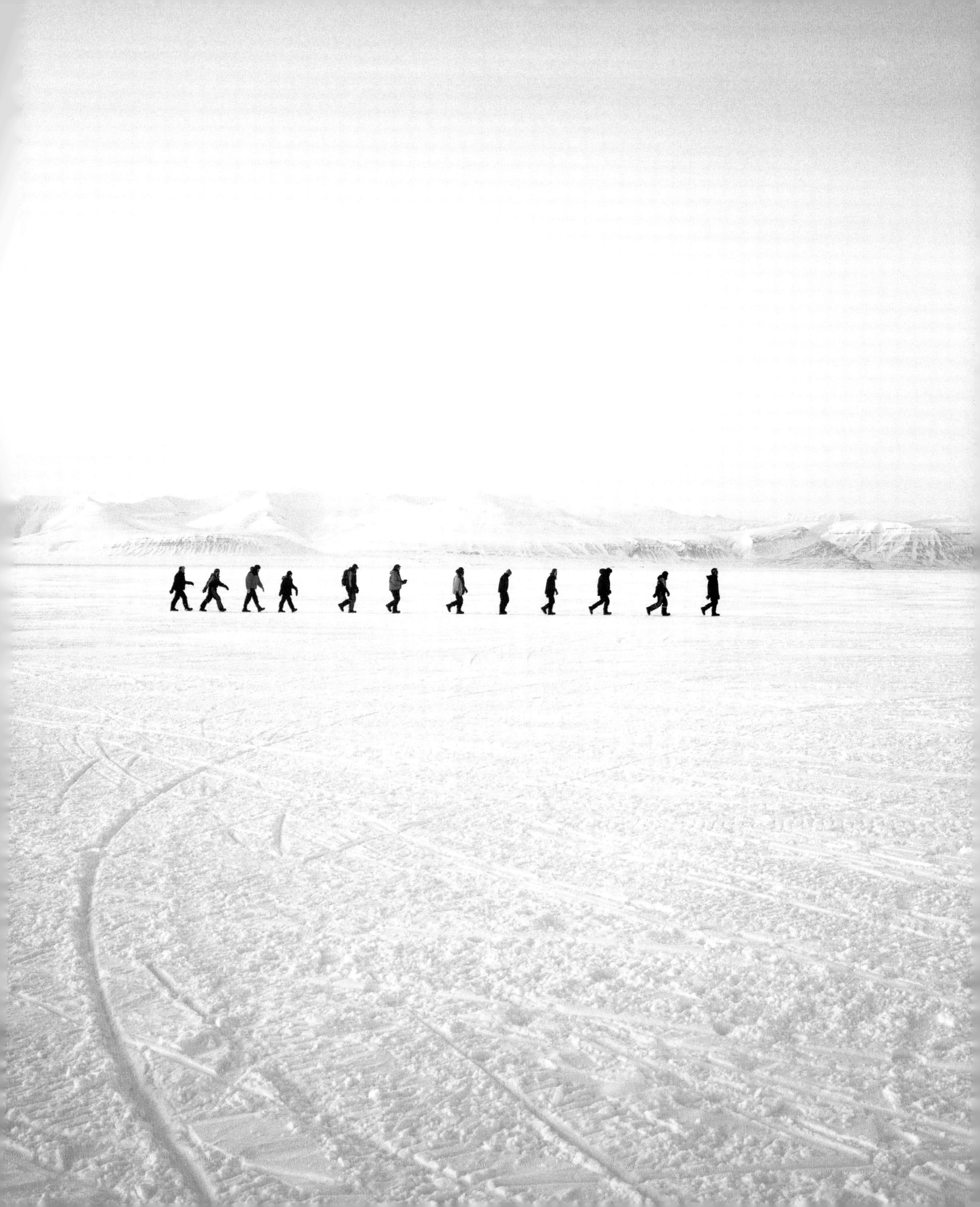

⊕ **Walking Dance** 2005/ 78°30N, 16°30E

'THE BEAUTY, SHEER PHYSICAL BEAUTY, IS SOMETHING THAT HAS IMPRINTED ITSELF UPON ME. BUT IT IS A SKIN OVER SOMETHING FAR MORE FEROCIOUS. I SENSE A COLD, I SENSE A VULNERABILITY. I FEEL MYSELF AS SOMETHING HOT AND BLOODY. MY BODY, IF IT WERE HARMED, THE FLESH WOULD BLEED. SO IF I FIND THE LITTLE BIT OF WARMTH I HAVE, I NEED TO PROTECT IT. THE IDEA OF PROTECTION, OF CARE, SEEMS PARTICULARLY MOMENTOUS HERE.'

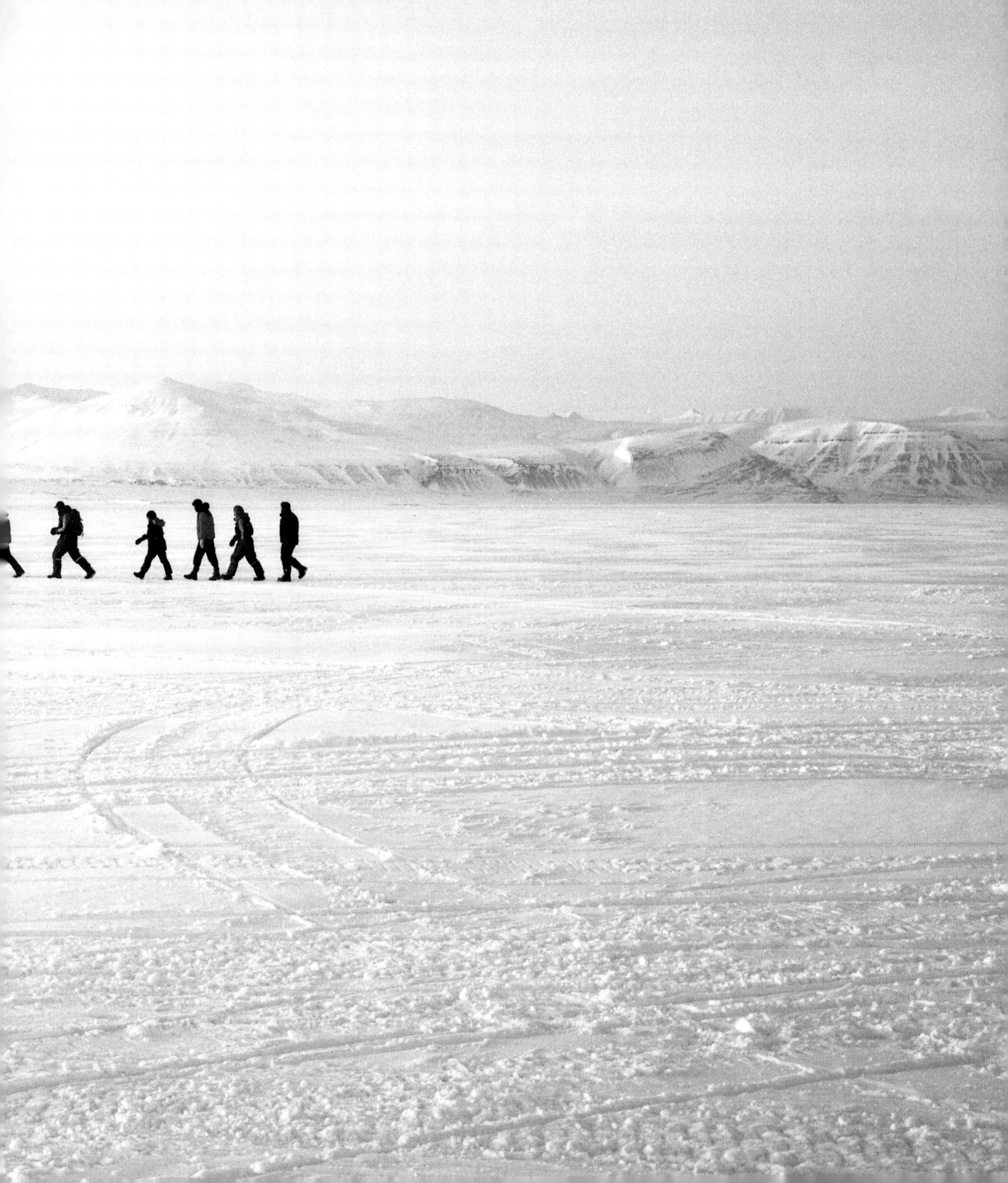

**Endangered Species** Greg Hilty, Director of PlusEquals Arts Consultancy

Climate change is not new news any more. Five years ago, when David Buckland initiated his series of artistic explorations of the changing climate under the banner of 'Cape Farewell', the phrase itself had only limited public recognition. People could shrug off its validity or urgency with relative nonchalance. That is scarcely an option today, deluged as we are with information and opinion on the subject. In a few short years the debate has shifted from weighing the evidence for the human causes of global warming to questioning how best to minimise and adapt to its likely environmental impact.

So has Cape Farewell already done its job, raising its highly distinctive voice alongside the grassroots and official campaigns to increase public awareness of climate change? If communication was its only aim then this might be the case. But Cape Farewell, in its ambitions and methods, has from the outset represented a more open exercise, a true expedition of untrammelled minds into a changing world of uncertain forms and feelings. Its outcomes were never pre-determined in methodology or propagandistic in intent. Its significance is consequently much greater.

At the heart of the Cape Farewell enterprise lay three expeditions to the High Arctic on board the schooner Noorderlicht in 2003, 2004 and 2005, undertaken by artists from disciplines including film, photography, installation, cartooning, sound, painting, sculpture, dance, architecture, digital media, and literature alongside scientists, educationalists and journalists. The act of travelling to this remote part of the world was itself essential. It took people with radically different world-perspectives to a place on the planet where climate change could be most clearly witnessed, both through scientific measurements of the Gulf Stream, on which our current climate depends, and simple observations, even over such a short space of time, of vanishing ice cap and shifting geography. The extreme cold and beauty of the Svalbard landscape represented a dramatic, sensorially and visually, bracing break from the participants' everyday working lives and personal preoccupations, cushioned by the supporting structures and technologies we take for granted in highly developed countries.

While the context of the expeditions was therefore charged with import, its focus was distinctly personal and its anticipated outcomes refreshingly undefined. The scientists had specific work to do, building on the existing body of knowledge relating to climate change. The artists had no defined brief to behave in any specific way or to deliver concrete outcomes. Even if they had had a general expectation of producing art objects, they had no standard methodology to bring to the context other than the history of their own creative practices, carried out in strikingly different surroundings. This unstable condition represented a concise simulation of the condition we will all find ourselves in, faced with some future radical environmental change.

What they did in this condition is revealing. Their physical survival having been more or less assured by the expedition organisers' careful planning, they used their time to try to understand the meanings that could be drawn from their unfamiliar new context. These 'meanings' consisted of knowledge acquired from research or observation, re-framed within each participant's own existing knowledge-base and mental maps. The characteristic activities they engaged in included collaboration with their colleagues and the creation of images.

Some of the activities of the Cape Farewell participants revealed structural and emotional similarities across different disciplines. An example of a scientifically-motivated activity was Dr. Simon Boxall's lowering of devices into the icy waters to measure the temperature or salinity of the seawater at varying depths to compare with the same measures taken over time. Max Eastley, sound artist, dropped sound recording instruments into the same water and came away with a sonic portrait of the Gulf Stream waters. Dan Harvey and Heather Ackroyd sank cubes of plaster, which the flowing waters sculpted into solid records of their movement patterns. None of these actions or resulting images was more 'real' than any other. Together they form a multi-dimensional representation that helps us both understand and emotionally grasp the dynamic nature of the sea off Svalbard.

Some of the image-collaborations were immediate and overt, some took longer and were manifested months later, when ideas generated on an expedition found form in an artist's work. In view of the limited materials to hand, it is remarkable how varied these artistic manifestations were. Sculptor Antony Gormley worked with architect Peter Clegg to define three spatial forms carved into the ice itself, each representing different aspects of the human body's relationship to the world — material, conscious, communal. Rachel Whiteread walked, talked and took photographs, over time absorbing the information of this alien environment into her longstanding artistic concerns, which would re-surface as one layer of meaning in her major sculptural installation in the Tate Modern Turbine Hall six months later. Alex Hartley engaged fully with the immediacy of the experience on all three expeditions, but abstracted his artistic manifestation in a long-term project to claim and name a small island whose existence was only revealed through the progressive retreat of the glacial ice between the first and second journeys. These works had radically different physical, spatial and temporal manifestations. They were all works of art, carefully-considered actions within the frame of contemporary sculpture; including landscape, conceptual and installation art where languages and formal motives derived from traditional sculptural practice are extended to apply to virtual, imaginative and emotional space and where context is included as if it were a material. All somehow graspable in the imagination of viewers, like myself, who were not present at their inception, these works also provide portals of comprehension for ideas and experiences relating to the impacts of climate change.

One artist's response to the Cape Farewell experience deserves a detailed look for what it says about the place of collaboration and image-making in the shaping of human consciousness. Choreographer Siobhan Davies went on the third voyage in March 2005. She went, she has said, anticipating feelings of wonder and awe in the face of the vast open spaces of the arctic. Once there, her overwhelming sensation was 'I must go for a walk.' Doing so took considerable preparation to ensure the body was padded and protected against the cold and the wind. The insouciant independence of the urban being was obliterated and Siobhan's overriding impressions were not so much of the landscape she moved through but the vulnerable, almost incidental body she had brought into this uncompromising environment. Accustomed through her profession to using her body expressively, she found expression here severely limited in nature and range. Her attention focused on her bones, skin, breath; on the effort and basic purpose behind her every movement.

Back home in London, Siobhan quickly formed the idea for a work that would embody some of the primal emotions and rational thoughts the journey had evoked for her. She wanted to create an image of a small, semi-human figure, displayed in a museum vitrine as if it were a branch of the human species that had either died out or was yet to evolve, or existed in some parallel world. She drew on a distinctive passage from one of her dance company's recent major works, titled 'Plants & Ghosts' in which dancer Sarah Warsop had started from a single simple movement, replicating it through a choreographic process evoking cellular growth. The development from movement to phrase to dance passage quite quickly brought in the need for simple props, light flexible rods with which Sarah extended her ability to reach out in all directions.

Adapting this material for her Cape Farewell piece required collaboration with people from a number of distinct disciplines, each performing a clear and complementary role under Siobhan's direction. Sarah Warsop took on the responsibility of substantially reworking her earlier role and performing it to camera. Deborah May, a film-maker who had worked closely with the company before, was involved in developing the image from the early stages, knowing she would edit it into its final shape. Sam Collins took responsibility for the presentation of the work in space, researching methods of projecting film to give the appearance of three-dimensional presence. New collaborators included Bergit Arends, curator at the Natural History Museum, and Dr. Chris Stringer, a distinguished human biologist based at the same institution, who both offered perspectives on evolution in relation to the intended meanings of the piece. Jonathan Saunders, a young designer building a strong reputation within the fashion industry, readily agreed to offer his help to create clothing for the figure that would enhance its capacity for self-extension into the space surrounding it.

These roles and relationships are worth spelling out because they were all independently and collectively crucial to developing, refining and realising the artist's ideas, first formed amid the frozen expanses of the High Arctic, into a coherent and powerful image capable of conveying those ideas to a wide public. The final image is a true hybrid of these different inputs, fit to survive only in its designated museum or gallery context.

A revealing moment occurred in Siobhan's first meeting with Jonathan Saunders, when he instantly recognised one historical inspiration for the 'movement creature' with sticks in the theatrical research that emerged from the Bauhaus, particularly the work of Oskar Schlemmer. The Bauhaus was founded in Weimar Germany after the trauma of the First World War and a long period of global cultural and economic strain. It sought renewal through design principles of usefulness and simplicity and collaboration across creative disciplines.

Schlemmer's observations in a 1927 lecture on the 'uses' of theatre ring startlingly true in the context of Cape Farewell:

*'We shall observe the appearance of the human figure as an event and recognise that at the very moment it has become a part of the stage, it is a 'space-bewitched' creature, so to speak. With a certainty that is automatic, each gesture and each movement is drawn into the sphere of significance …*

*From this point on, two fundamentally different creative paths are possible: that of psychic expression, of emotional passion, of dramatic mimetics and gesture; or that of motion-mathematics, of the mechanics of joints and numerical rhythmics and gymnastics. Each of these paths, if pursued to the end, can result in the highest achievement, just as their fusion can lead to a unified art form. The actor is now so susceptible to being altered, transformed, or bewitched by each of the objects applied to him – mask, costume, prop – that his habitual behaviour, his physical and psychic structure, are thrown off or put into a different balance … '*[1]

Recent evolutionary theories draw on new evidence of the quite small number of genes that make up even sophisticated species like humans and the relatively high number of genes common across the spectrum of simple to complex organisms. It is argued that evolutionary adaptations are not hardwired into our genetic makeup but 'switched' on or 'expressed' in relation to external circumstances. Without making too much of metaphorical comparisons, we can surely find emotional resonance and value in following Schlemmer's image of humans as all 'space-bewitched', all more or less consciously acting out our biologies and our destinies within the environment we inhabit.

Art and the sciences represent parallel, mutually-reinforcing acts of feeling our way through our environment, understanding and asserting our presence as a means of preserving it. Thus the collaborative research and image-making represented by Cape Farewell should surely be the norm rather than the exception, standard practice rather than a romantic exception. This is not to say that Siobhan Davies, or any of the other artists of Cape Farewell, would pretend to have answers to the major problems posed by the reality of climate change. The individual artwork does not have to be directly 'instrumental' to be 'useful'. The image-making role itself is fundamental to human survival, such as sleeping, or having a sense of direction, or more complex behaviours like instinctive altruism. Artistic exploration is a human activity as old, widespread and diverse as technological innovation on the real stage of human activity.

Antonio Damasio has argued persuasively for the centrality of image-making to supporting and shaping consciousness. At the most basic level, he asserts that: 'Life and the life urge inside the boundary that circumscribes an organism precede the appearance of nervous systems, of brains. But when brains appear on the scene, they are still about life, and they do preserve and expand the ability to sense the internal state, to hold know-how in dispositions, and to use those dispositions to respond to changes in the environment that surrounds brains.'[2]

Climate change offers huge challenges to our societies, not just as technical problems but as major tests of our capacity for collaboration and imagination. Yet its challenges are finally so basic that it reminds those who continue to find comfort in the self-contained worlds of business, or science, or the arts, that their disciplines however refined are 'still about life'. For our species to survive and flourish we will need to exercise our brains across multiple dispositions.

[1] Oskar Schlemmer, 'Buhne (Stage)' lecture delivered March 16 1927 in Dessau, published in 'The Bauhaus Journal' 1927, No. 3, reproduced p474 Hans M. Wingler 'The Bauhaus' MIT Press, Massachusetts, 1969, 1978

[2] p139 Antonio Damasio, 'The Feeling of What Happens' Vintage, London 1999, 2000

**Endangered Species** 2006, 51°31N 1°95W

**Journal, 20/09/2004** David Buckland

We walked the two kilometres to shore yesterday evening, conversations drifting as we tried to get a measure of just how large the scale of this place is — wonderful to attempt to comprehend, impossible to represent. Every time you move, you take another photograph, and after every photograph, the light changes and you photograph it again. But there is something about all the messages that are in my head that are not coming through in the photographs. What the artist does is to somehow find the human scale of the storytelling. In this case, the human scale of climate change.
My ambition was to find a way of including the additional narrative of the climate debate within my artwork from the Arctic.

5a.m. and -10°C on the 79th parallel north, still dark. We had our projector mounted on the gunnels, 20 foot from a towering wall of ice, as high as the ship's mast. We were poised to project my texts, photograph the result and film the process. I had prepared 16 slogans from the writings of Gretel Ehrlich in her book The Future of Ice from the first Cape Farewell expedition. At the cusp of dawn, the projection matched the intensity of the daylight on the glacier front. I am still in awe of Gert our captain and Maaike the first mate, who at times sailed so close to the ice wall we could almost touch it. Mercifully it was calm and the glacier relatively dormant, not calving. The projections pierced the icy surface of the glacier. At times the projection disappeared into the glacier and at others it reflected clearly, it was amazing to discover so many different kinds of ice.

**The Cold Library of Ice** 2004 / 78°55N, 19°05E

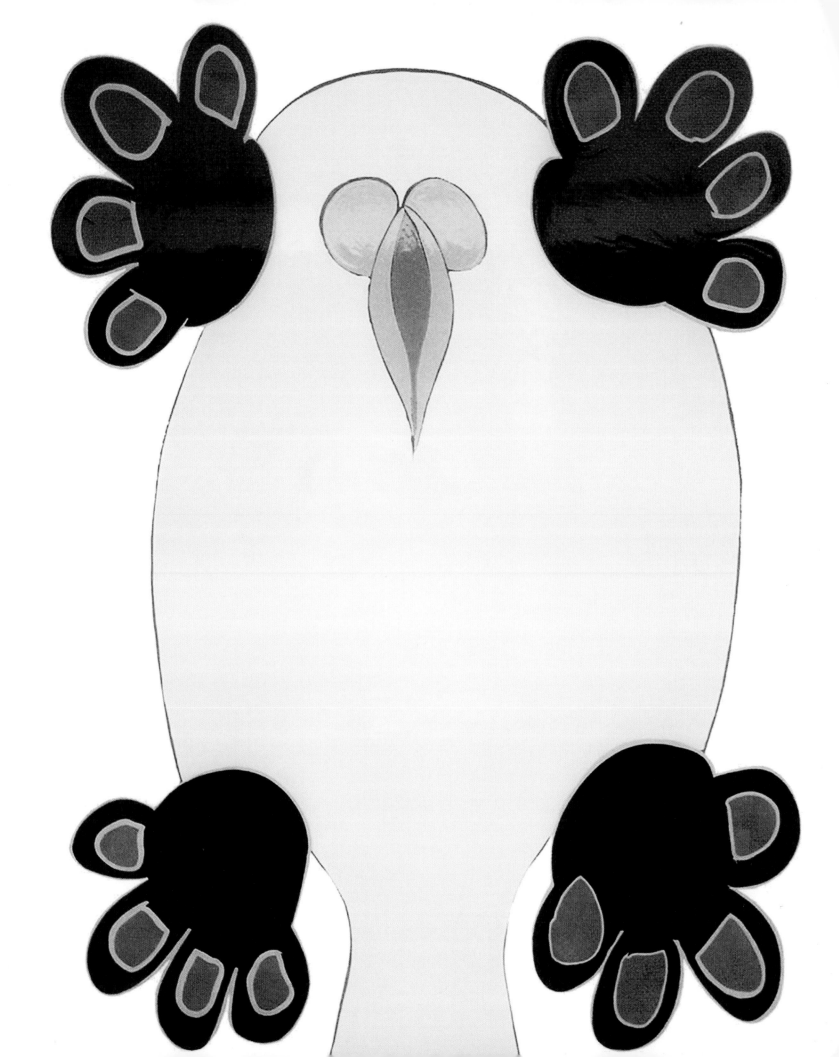

### Hermaphrodite Polar Bears

It has recently been discovered that Arctic polar bears are being poisoned by certain chemical compounds, commonly used in Europe and North America. Significant deposits of these polybrominated diphenyls or PBDEs, used in flame-retardants for household furnishings, have been found in the fatty tissue of polar bears, especially in eastern Greenland and the Svalbard Islands. The team of scientists who are carrying out studies on the effects of PBDEs on polar bears are concerned because in tests these chemical compounds attack the sex and thyroid glands, motor skills and brain function of laboratory animals. Evidence also suggests that compounds similar to PBDEs have contributed to the astonishingly high rate of hermaphroditism in polar bears. Around one in fifty female bears in Svalbard displays both male and female sex organs.

The Arctic is suffering from high levels of pollution from man-made compounds, such as flame-retardants. Chemicals we use everyday in the home in Europe and North America are migrating northwards in winds and currents. Contaminated moisture condenses on arrival at the Arctic and enters the aquatic food web. The concentration of the compounds biomagnifies as it ascends the food chain from plankton to fish to marine mammals such as seals, whales and polar bears. A study published in Environmental Science and Technology in December 2005[1] showed that one compound was 71 times more concentrated in polar bears than in the seals they eat. The results of this new research are particularly disquieting as the Arctic polar bear population is already critically threatened by the destruction of its natural habitat and hunting grounds due to global warming.

[1] Derek C.G. Muir, Sean Backus, Andrew E. Derocher, Rune Dietz, Thomas J. Evans, Geir W. Gabrielsen, John Nagy, Ross J. Norstrum, Christian Sonne, Ian Stirling, Mitch K. Taylor and Robert J. Letcher, 'Brominated Flame Retardants in Polar Bears (Ursus maritimus) from Alaska, the Canadian Arctic, East Greenland, and Svalbard' in 'Environmental Science and Technology' 40 (2) December 2006 pp449-455

**Hermaphrodite Polar Bear** Gary Hume, Artist

##  UK Cetacean Strandings Project  Richard C. Sabin, Curator, Natural History Museum

As a curator in the Natural History Museum's Department of Zoology, I work with one of the best mammal research collections in the world. It is a sobering experience to find yourself handling the bones of animals brought to the verge of extinction by human interaction, whether through hunting, habitat destruction, pollution or over-exploitation of resources. Many of these issues were championed in the twentieth century by conservation groups and governments across the world, and led to a huge increase in public awareness about the fragility and complexity of global ecosystems. Meanwhile, in the background something bigger was brewing, something that had been acknowledged but whose speed and effects had been wildly underestimated. With global warming posing such a big threat, have our concerns about the planet over the past 100 years been in vain? Have our collective efforts been a waste of time? The answer I feel, is no. They are more important now than they ever were.

One of my main areas of responsibility is to manage the National History Museum's UK Cetacean Strandings Project, set up in 1913. We carry out biogeographical monitoring of whales, dolphins and porpoises (collectively known as Cetacea) around our coastline. We explore their biology, diet and species distribution, seasonality and frequency. If a cetacean is found stranded on a UK beach, a post mortem examination is carried out by one of our project partners such as the Zoological Society of London. This is to find out why it died, how old it was and what diseases it may have been suffering from. We have recorded an impressive 25 different species of cetaceans in the past 93 years, almost one third of those known worldwide. We have worked hard to pass on our data to the public and let them know just how rich the marine life is in our small corner of the globe. Through this project, I came to know the artists Heather Ackroyd & Dan Harvey, who wanted to create a crystal-encrusted whale. I was able to help them secure the skeleton of a stranded minke whale, one that had been found dead near Skegness, on the east coast of England. The minke whale is the smallest and most common of the baleen, or filter-feeding, whales that live around the UK coastline. On average, each year we see five to 20 strandings of this species, most of which occur around the northern and western coasts of Scotland.

The strandings project has taught us much about the seasonal movements of marine animals around the UK over the years, allowing us to identify which species we expect to see, where and at what time of the year. In the past 40 years, we have begun to see species previously unknown in UK waters, such as the pygmy sperm whale (first stranding record, 1966) and Fraser's dolphin (first stranding record, 1996). Other more familiar species appear to be changing their distribution, like the striped dolphin, white-beaked dolphin and short-beaked common dolphin. Some of our colleagues believe these changes are due to the effects of climate change, while others feel not enough evidence exists to make this claim. It is also possible we are seeing changes which are the result of naturally occurring long-term processes which cycle through the world's oceans altering sea temperatures and the distribution of both predator and prey animals.

The possible effects of climate change on marine animals, particularly cetaceans, are difficult to see because of the 'hidden' nature of the environment in which they live. Beneath the waves, changes may be taking place. At an international scientific conference hosted by the UK Meteorological Office in February 2005, it was reported that large amounts of fresh water pouring into the sea from melting glaciers could have a dramatic effect on the transfer of heat from the equator to higher latitudes. This could mean a cooling of the world's oceans, which would be bad news for many species of cetaceans.

The effects of climate change on land mammals will be more immediately obvious. We are already seeing dramatic images in the Media, showing the melting of the permafrost in sub-Arctic regions. An unexpected by-product of this process has been the glut of available mammoth ivory from the Siberian Arctic, released from the icy grip of the frozen earth after thousands of years. This material is finding its way onto the commercial ivory market, and as a result, my colleagues and I here at the Natural History Museum have seen an increase in the amount of seized material brought to us for examination by Her Majesty's Customs and Excise.

The large land mammals of the Arctic and sub-Arctic region, like the musk ox, the polar bear and the caribou, are all highly specialised and very well adapted to their environment. Left unchecked, the potential rapid rate of climate change, which has been forecast by scientific bodies around the world, may alter the environment faster than these animals are able to adapt to it, possibly leading to their extinction.

It is difficult for me to imagine a world in which the animals I have come to love are absent. How are we going to convey to our offspring, the awe, the joy, the excitement and even the fear at seeing wild animals in their natural environment, if both the animals and their environments have vanished? Despite the many cultural differences of our species, I think we now have one thing in common with each other – the prospect of universal loss.

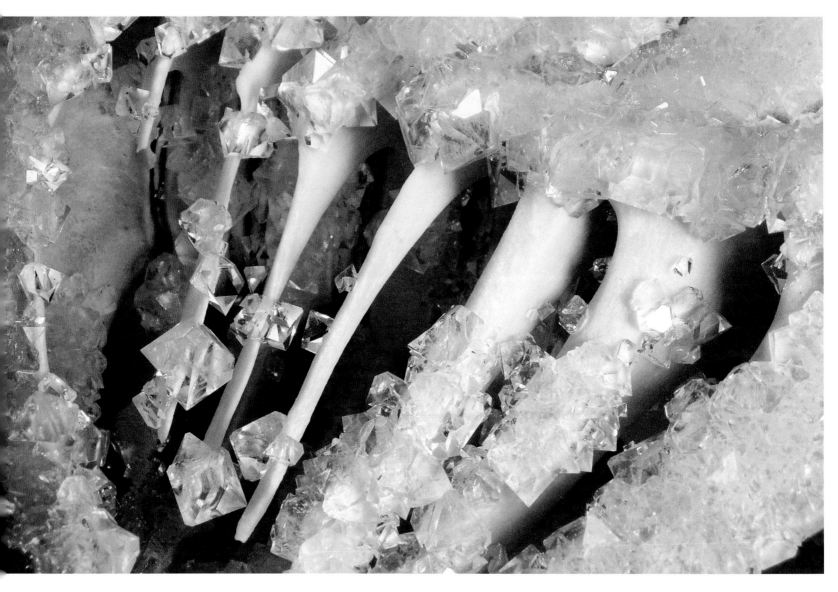

⊕ **Whale** 2004 / 80°05N, 14°58E

'ON THE FIRST TRIP UP TO SVALBARD I CAME ACROSS PICTURES OF BEACHES THAT WERE COMPLETELY COVERED IN BONES. ALREADY HAVING A FASCINATION WITH BONES, IT WAS SOMETHING I WANTED TO SEE AND EXPERIENCE.'

## Stranded  Heather Ackroyd & Dan Harvey, Artists

18 October 2005: Richard Sabin from the Natural History Museum calls with news of a minke whale washed up on the beach near Skegness, email images arrive, it seems good, skeleton intact. We start to organise ourselves, collecting together our whale flensing kit: knives, sharpeners, plastic bags and sheets, rubber gloves, overalls, rope and numbering tags. The whale is resting almost half-buried on a stretch of sandy beach called Gibraltar Point. The council provides a large skip for the flesh to be dumped in and we are ready to start cutting by mid-afternoon. On that first day we remove the two jawbones and separate the skull from the rest of the body. There is a light breeze that keeps the smell almost bearable; the six-metre carcass has been dead for approximately a month judging by the state of its decomposition.

There is quite a precise way that one needs to approach removing the bones of a whale, first taking off the head and cutting off as much meat as possible, then working along the backbone, cutting halfway down the ribs and letting the mass of the flesh peel back by its own weight. The knives blunt fast and need constant sharpening. The ribs, in fact all the bones, are covered with a thick sheaf of almost golden skin, which is firmly attached but can be removed like a stocking once cut away. The ribs protect all the internal organs and Richard has forewarned us to be careful of exposure to pathogens when cutting into the lungs. The stomach and intestines are easy to spot but with the state of decomposition it is hard to tell apart the other organs. In fact, we do not think the lungs are there. When a whale dies its internal temperature will begin to rise as heat is generated from the rotting process, the blubber acts as a very good insulator and as gases build up inside the corpse, the lungs are violently expelled through the mouth. Each bone is bagged, numbered, sealed and tagged on the outside. The backbone is separated into sections of four or five vertebrae. After two and a half days we have a skip with about five tons of putrid flesh in it and a pile of plastic-wrapped, smelly bones. It is a very particular, pungent smell of death, one that enters your nose and haunts you.

We are under the impression that the smell could not get any worse, but leaving bones wrapped in plastic for the best part of a week is another thing. Although the bags are fairly well sealed, blood seeps out and flies slip in. Then the hatching of maggots soon turns much of the remaining flesh to a grey soup. Weeks of boiling and cutting ensue, along with keeping track of which bones go where. Each vertebra has two end discs (biscuits) that in an older creature would be fused to the bone. We realise that the minke, whilst fully grown, was still an adolescent. Stripping this creature right down to his bones, we become connected to him in an intimate and intense way, as if his death is giving us a closer understanding of his life.

The final cleaning process is done with enzymes. The bones no longer smell and almost look as if they are carved. A perfect specimen ready for the next stage: crystallisation. We begin with the bones weighted down to base plates, as they lie for days in dark, steaming tanks of a supersaturated chemical solution. As the solution cools a reaction happens; seeds form, attach themselves to the bones and begin to grow. Slowly the skeleton is encrusted with crystals, a strange embalming in chemical water. As we draw to the surface the crystal covered bones, they are exquisitely fragile and any sharp movement will cause crystals to fall.

Ephemeral is a word we often use in describing our work. Our site-specific artwork involves processes of growth, decay, erosion and transformation. It is transient. It is about change. And in our world there is continual change, and therefore continual loss. So how and why does loss matter?

The chemistry of our oceans is changing. It is now accepted that, if we continue unabated in our consumption of fossil fuel, the acidity of the oceans will increase incrementally and the life they support, from tiny carbonate shells and plankton upwards, will perish.

Does the loss of this young minke whale matter? We only know him through his death and we really know very little of his life. So we have embraced his loss and sought to reappraise the value of his life. Here, where the gleaming crystals encase his lifeless bones, lies a memento mori for our times.

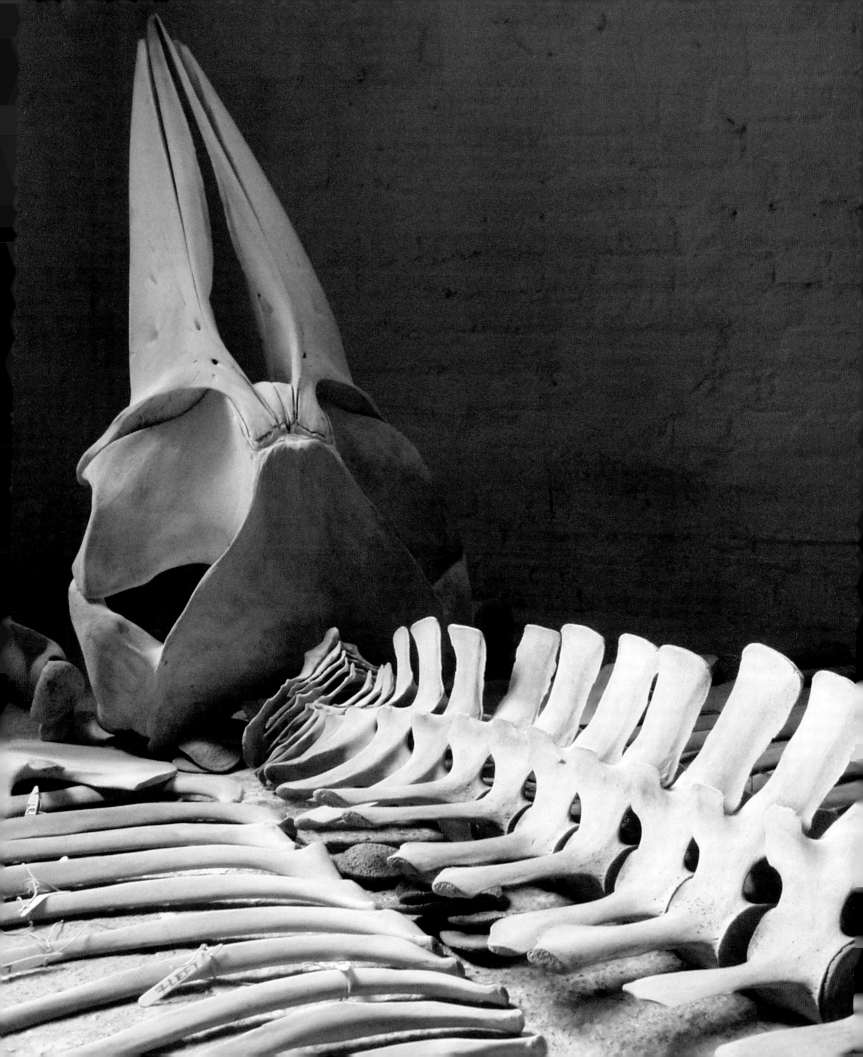

**Facing Climate Change: Tipping Points and U-Turns** Prof. H. J. Schellnhuber, Director, Potsdam Institute for Climate Impact Research

Working as a climate change scientist means that your everyday professional life is dominated by abstract representations, such as equations, tables and figures. Yet 'anthropogenic interference with the climate system' is not a computer phenomenon played out in cyberspace: global warming is happening already and its impact can be seen, smelled and touched in thousands of places around the globe. I think it is imperative to keep acknowledging its geographical imminence and indisputability while we steadfastly try to advance our understanding of the process through sophisticated research and computer technology. Never must we lose the sense of what is at stake when we gamble with our planet.

Travelling as a member of the Intergovernmental Panel on Climate Change (IPCC) group of assessment experts offers ample opportunity to discover global environmental change through firsthand experience. For a number of reasons, not least regional representation within the UN system, the venues for IPCC meetings include several distant, exotic and utterly beautiful locations. But this beauty often seems doomed to vanish. For example, one of the recent IPCC assemblies took place in Mèrida on the Yucatán peninsula, which was devastated, just a few weeks earlier, by a powerful hurricane. With rising sea-surface temperatures in the wake of manmade atmospheric warming, the destructive energy of tropical storms is bound to increase significantly. Therefore, it is doubtful whether the Caribbean, where even grave political mismanagement has not yet been able to break the spirit of her peoples, can continue to subsist as a haven of serenity.

Back in 1999, I participated in an IPCC meeting in Canberra, the Australian capital. These were unusually cool December days at the beginning of the Southern Hemisphere summer, filled with hard work and exhausting discussions. Towards the end of the gathering, however, we were invited to tour the nearby Tidbinbilla National Park, which was teeming at that time with wildlife, especially koala bears. In February 2006, I revisited South East Australia and was struck by the dramatic changes that had taken place there.

In 2003, Tidbinbilla was devastated by a firestorm that turned day into night in the nearby capital. Only one single koala survived with heavy injuries because he had been idling in a creek when apocalypse descended. 'Lucky', as the park rangers dubbed him, went on display in the visitor centre as a touching example of a miraculous escape. Sadly, however, when I went to witness the miracle, a plaque announced 'Lucky is retired'. In early 2006 an unprecedented heat wave, with temperatures soaring far above 40° C, afflicted Sydney and Melbourne. Then on 20th March 2006, the category 5 cyclone 'Larry' crashed into Queensland, destroying animal habitats, harvests and infrastructures but, fortunately, no human lives. This event may be perceived as a Southern Hemisphere comment on the 2005 hurricane season in the Caribbean that broke all historical records. It will also increase the concerns of Australian environmentalists about the future of one of the natural wonders of our world, the Great Barrier Reef.

The biggest and most magnificent marine ecosystem on Earth is threatened by multiple stresses that are related to anthropogenic $CO_2$ emissions. The most direct impact stems from ocean acidification: as seawater turns sour, the calcification processes in coral reefs are modified and partially suppressed. This exacerbates coral bleaching dynamics, triggered by higher sea-surface temperatures as a more indirect effect of $CO_2$ accumulation in the atmosphere. Scientific investigations in the Caribbean now indicate that coral reefs suffer tremendously from tropical storms and tend to recover only very slowly after the hit. Therefore, an intensified cyclone regime in the North East Australian seas could prove to be the last nail in the coffin of the Great Barrier Reef. By 2050, this jewel in the crown of our world heritage may have degraded beyond repair.

This tale of the 'other' hemisphere can teach us a number of lessons. One of the crucial ones is that unabated global change will move nature and society towards and beyond 'tipping points', where unique systems start to collapse in an irreversible manner – on a human time scale, at least. There is probably no such thing as a 'planetary tipping

# 'THE GREENLAND MELTDOWN ALONE WOULD RAISE GLOBAL SEA LEVEL BY 7 METRES WITHIN THIS MILLENNIUM.'

point', that is a specific increase in global mean temperature that could trigger a run-away greenhouse effect as a self-amplifying, non-linear process. However, there is a growing body of evidence indicating that an entire cluster of critical thresholds lurks in the Earth System waiting to be transgressed, one by one, as humankind turns up the heat.

Prominent examples of major elements in the planetary machinery that might be tipped into fundamentally different states, or modes of operation, are: the Greenland and West Antarctic ice sheets; the Atlantic ocean currents, responsible for the benign climate across the British Isles; the El Niño-Southern Oscillation system; the Asian and African monsoons; the methane reservoirs on and off various continents; the Amazon rainforest and the boreal forests, and the marine biological pump, that acts as a sink for excess $CO_2$. Apart from the direct impacts of such switching processes, the Greenland meltdown alone would raise global sea level by 7 metres within this millennium. Many of the phenomena listed are bound to interact through domino dynamics and to accelerate anthropogenic climate change through positive feedback loops. The probability that this disturbing scenario will become reality increases significantly if the world warms more than 2° C above pre-industrial levels.

The consequences for human civilisation would be practically unmanageable. Even under the present climate conditions, and associated extreme event regimes, many societies operate on the brink of failure. They are doomed to collapse if afflicted by shocks of the type sketched above – just think of the impacts of a thoroughly transformed Indian summer monsoon. Furthermore, the capacity of industrialised countries in the mid-latitudes is not unlimited either, although the stresses of abrupt climate change might not be primarily a matter of life and death in those parts of the globe. In addition, the 'developed world' will have to digest secondary shockwaves as generated by the implosion of the most vulnerable societies in the 'developing world'.

At a recent British-German workshop sponsored by the Foreign & Commonwealth Office, leading climate change scientists from both countries tried for the first time to envisage the mechanisms driving such a state of crisis, and to short-list major socio-economic systems that are prone to be 'tipped' in the course of certain events. Apart from identifying a number of regions at high risk, including rapidly urbanising areas in semi-arid and coastal zones, the participants zoomed in on three global systems: fisheries, public health infrastructure, and international refugee management. Concerted research action is urgently needed to bring our understanding of the stability of socio-economic systems up to an operational level.

Most potential disruptions may be avoided if the global warming threshold of 2° C above the pre-industrial value can be sustained. This stabilisation of planetary temperature, however, cannot be achieved by simply stabilising our unsustainable behaviour. In particular, the total $CO_2$ emissions will need to peak soon and decrease afterwards to ensure that the atmospheric concentration of this greenhouse gas bends away from 450 ppm level. This means, in fact, that our society has to prepare and implement a number of u-turns affecting many aspects of modern life such as production, consumption, housing and mobility. Economic analysis with advanced energy-climate models demonstrates convincingly that we can afford this decarbonisation of our civilisation. The question remains whether we can deliver it in time.

Multiple natural and social inertias keep moving us closer and closer to various tipping points, like a boat floating on a fast-moving river towards a waterfall. To stop rowing forwards is not enough, we need to turn around and head upstream! Public awareness of the severity and urgency of the climate crisis is necessary for such a u-turn for sustainability to be made. Scientists and artists will be able to bring about that awareness if they are willing to work together over the coming decade.

**Journal 12/09/2005** Heather Ackroyd

The Noorderlicht lay captive in a frozen bed of sea, a wonderful vessel full of warmth, food, light and everything we needed to sustain us on our brief stay in this frozen land. We arrived in a boot-lace convoy of skidoos, the hyper-active drone like demented mosquitoes, rupturing the tranquility. Don't get us going on skidoos, or skidon'ts, as they were rapidly termed, as they variously refused to start whilst we stamped our feet and breathed in clouds of fumes. It would take 6 huskies a short day to bring a pair of travelers here to the boat – a journey the skidoo does in two flat hours. Time is of the essence here and we don't seem to move at a dog's pace anymore.

Dan and I carved a lump of glacial ice we sawed and hauled out of the ice-sea into a large disc-shaped lens. We'd hoped it would focus beams of sunlight, possibly melting snow or scorching sheets of paper. Not to be. The sun was low and weak in the sky but the glacial lens became a sun-catcher presenting a cracked mosaic of light. Dan made a camera using a slice of ice, placing it behind a lens in a block of snow. The image of sun and distant mountains appeared like an apparition, inverted on a sheet of ice.

**Journal 12/09/2005** Dan Harvey

It is often the unexpected that can be illuminating and lead you to other processes. Trying to stay true to one's location and allow it to dictate the nature of the work is not easy at minus 30, but it is exciting – like working on another planet.

The word 'frozen' makes one think of something static. This is not so. Even at these temperatures the glaciers are still moving, shedding their slow load; the solid sea buckles at the shoreline as it expands, pushing up or down and cracking, releasing liquid water that then in turn, turns solid. What does play on my mind after this incredible experience is the impact that skidoos are having on this environment. Is it true that one hour on a skidoo produces the same amount of hydrocarbons as 10,000 kilometres in a modern car with a catalytic converter? It often comes down to speed, the 'quick fix'. At one point whilst cutting ice, we used a large handsaw but as evening fell and with freezing hands the option of using a chainsaw seemed to make sense. I have often used a chainsaw, but up here suddenly one questions it. Like many modern tools that make our lives easier, they take their toll here. Eco tourism: Can there really be such a thing?

Ice Lens 2005 / 78°30N, 16°10E

**⊕ Ice Lens** 2005 / 78°30N, 16°10E

'WE'D HOPED IT WOULD FOCUS BEAMS OF SUNLIGHT, POSSIBLY MELTING SNOW OR SCORCHING SHEETS OF PAPER. NOT TO BE. THE SUN WAS LOW AND WEAK IN THE SKY BUT THE GLACIAL LENS BECAME A SUN-CATCHER PRESENTING A CRACKED MOSAIC OF LIGHT.'

**Ice Slides** 2005 / 78°30N, 16°10E

⊕ **Marker** 2005 / 78°30N, 16°10E

**The Science of Climate Change** Sir David King, Chief Scientific Adviser to the H. M. Government

The world is getting warmer. Over the past century the global climate has warmed by an average of 0.7°C, with much of this seen over the past 30 years; in fact 19 of the hottest 20 years in the past 150 have occurred since 1980. The effects of this warming are visible. Ice sheets are melting, sea levels are rising, and glaciers across the world are in retreat. The impact on human life is already being felt. The 2003 European heat wave is estimated to have caused over 30,000 premature deaths, with an estimated direct economic cost of $13.5bn.

The causal link between global warming and increased greenhouse gas emissions from human activities, most notably carbon dioxide, is now established beyond all reasonable doubt. While, even a year ago, climate change was still reported as a controversial issue, new scientific findings have been filling in the gaps in our understanding and have countered the last of the sceptics' arguments. The good news is that we now understand what is happening, and why. We know that to avoid the worst impacts of climate change we must reduce our greenhouse gas emissions, of which carbon dioxide is the biggest contributor.

Climate science is now a mature subject and the basic science on which it rests is well understood. This enables us to understand the much needed and naturally occurring 'greenhouse effect'. Energy from the Sun warms the surface of the Earth, which in turn radiates heat back into the atmosphere. Gases such as carbon dioxide, water vapour and methane – the so-called greenhouse gases – absorb some of this radiated heat and act like a blanket, warming the Earth to an average temperature of about 15°C. Without the greenhouse effect life on this planet would not exist as we know it, as the average temperature would be a chilly -18°C. In this way, greenhouse gases, in particular carbon dioxide, have a benign effect in keeping our planet warmer and its temperature more stable than it would otherwise be.

However, increasing amounts of carbon dioxide and other greenhouse gases in the atmosphere have been shown to be causing the planet's temperature to rise. Carbon dioxide levels are rising fast. The graph shows the carbon dioxide levels in the atmosphere over the last 30,000 years. The atmospheric concentration has risen by about 40% since the Industrial Revolution began in Europe in the middle of the 18th century, and the current rate of increase is the highest on record. The levels of carbon dioxide now in our atmosphere have not been seen for at least 650,000 years, and probably more than 20 million years.

Scientists are able to bring much of our knowledge and understanding of the behaviour of the planet's atmosphere and oceans together to make computer simulations of the climate. The observed rise in temperature cannot be explained if only natural factors are included in the simulation. The rise can only be explained by adding in the effect of human activity. This activity includes our use of fossil fuels and our changing land use. Other direct evidence supports this view. The models used to explain current climate changes can also be used to predict our future climate. While there are uncertainties in predicting future human activity and in the finer detail of the science, there is a strong consensus that global warming will continue, leading

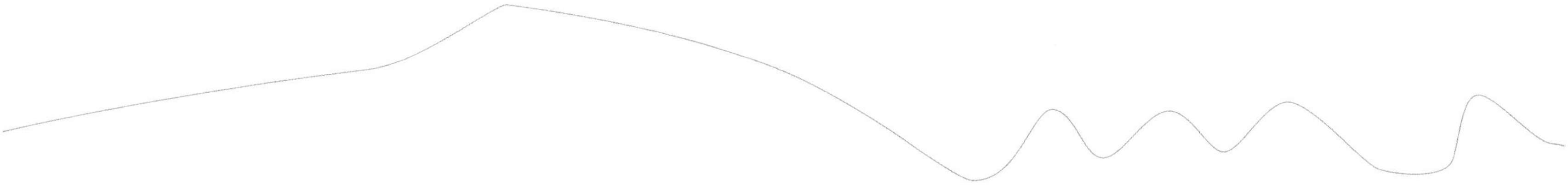

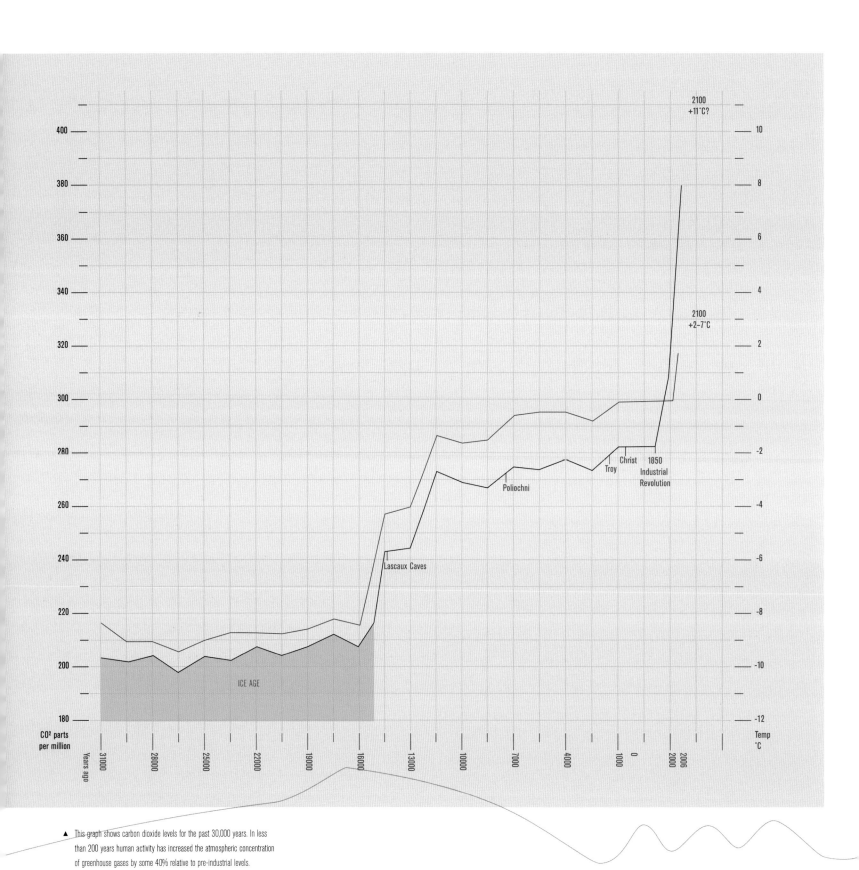

▲ This graph shows carbon dioxide levels for the past 30,000 years. In less than 200 years human activity has increased the atmospheric concentration of greenhouse gases by some 40% relative to pre-industrial levels.

to increases in the range of 2–6°C above 1990 levels by 2100. Some recent findings suggest an even higher figure of up to 11.5°C. The implications for our societies, and indeed the very lives of people across the planet, will be profound.

As global temperatures rise, we can expect increasing extremes in our weather. Recent experience in the UK and the rest of Europe, such as severe floods in 2002 as well as the 2003 heat wave, shows that extreme weather related events have significant human and economic costs. Higher temperatures will have knock on effects for climate across the globe, influenced strongly by regional factors, including more intense precipitation events, increased droughts and floods in tropical regions, and increased hurricane and typhoon wind intensity. The worst fallout from global warming will be experienced in the world's poorest countries, which are both the most vulnerable and also the least able to adapt.

A 2°C average global temperature rise, which is possible by the middle of the century, could mean as much as a 4°C rise in the middle of large continents like Africa. In a continent already ravaged by poverty, famine and disease, climate change is likely to worsen food and water security, accelerate irreversible losses of biodiversity and bring greater threats to human health.

Asia too is at severe risk, with the south east particularly vulnerable to rising sea level. For Bangladesh, a one metre rise in sea level would mean a loss of one-fifth of the viable land area, affecting 15 million people. For India, a similar rise could mean around 7 million people displaced. Climate impacts will also affect water and food security in the region. Those people living in low lying, smaller island states may be forced to abandon their environment altogether.

Climate change will intensify the effects of poverty in the world's developing regions through losses of biodiversity and agriculture, with adverse impacts on health affecting almost every sector of society. It takes little imagination to appreciate how such impacts would reverberate far beyond the regions most directly affected.

It is obvious that we must act, and quickly, both to reduce the future impacts of climate change and to adapt to those impacts that cannot be avoided. Climate change has no regard for international boundaries and every part of the world will be affected, though the impact will differ in different regions. It is a truly global problem that requires truly global solutions. And the earlier we start the more options we will have, the lower the risks and costs and the greater the chance of success.

This global action has already begun. An international convention, the United Nations Framework Convention on Climate Change, was first adopted in 1992 and has now been signed by 189 countries. It sets an ultimate objective of stabilising greenhouse gas emissions at a level that would prevent human activity dangerously interfering with the climate system. A key part of this framework is the Kyoto Protocol, which imposes legally binding targets on those developed countries that have ratified it. Different countries have different targets, which aim to reduce their overall emissions of a basket of six greenhouse gases by 5.2% below 1990 levels over the period 2008-2012. However, significant programmes of action are required to address this.

Governments will always be an important part of the solution as they set the framework within which citizens and businesses take decisions. But climate change is not only about what governments can do. Others need to play their full part too. Everybody is, in some way, part of the problem and can and must become part of the solution.

As a society, we need to come up with new and innovative ways to address the challenges and seize the opportunities. To produce and consume in smarter ways, uncover new, more sustainable ways of growing and boosting economic activity while accelerating the changeover to carbon neutrality. We need new and alternative sources of energy, better ways of working with the Earth's resources, more efficient transport of people and of goods and a more inclusive global society.

As citizens, we need to feel empowered to make the changes to our day-to-day lives that have an impact. Insulating our homes, turning off the lights, walking or cycling instead of getting in the car are just a few examples of the many simple actions we can take. They do not require us to reduce our quality of life, may be beneficial to our health or save money, and do help prevent climate change. The results of our actions will not be immediately apparent, just as the consequences of the emissions we produce now are not, but will certainly benefit our children and grandchildren.

I believe climate change is the biggest single global challenge that we face. Our success or failure in taking the steps necessary to tackle it now, and over the next couple of decades, will play out for centuries to come. If unchecked, and if we fail to adapt, it has the potential to be catastrophic. This is a serious problem and solving it will not be easy. But, with commitment and innovation, it can be done.

📄 **Meanwhile Back in Gotham** Michèle Noach, Artoonist

Tripping through the scalding attrition of Leicester Square yesterday, an inner voice had one question over and over. Was that pale blue dream true? How to remember?

Gazing in disbelief at the colossus of the Arctic, the kilometres of zinc white ice struck through with inexplicable cobalt and black. Semi-mythical narwhal swanking past in the ultramarine death pool our tiny boat floats upon. Climbing an ice-cap hued like the air, trying to remember to stay on earth, to not walk into the sky. Skinny polar bears appearing to greet self-sacrificing reindeer with a mimed peck on the cheek (did we really see that?). The Northern Lights peeling back the routine of night with a Haight-Ashbury display of sulky luminescence that seemed to mark out the exact dome of the sky; the top of it all is here, it said.

How to square these things with what is now?
The unavoidable spillover from what the oceanographers were busy observing with their 'scopes and graphs and screens and vials. These fragments of the real world, slivers that explain the real world, are also filtering through to how the whole is calibrated, understood and recalled.

This was no mini-break. The violent spirit of this voyage demands attention. The High Arctic is a gifted child that needs particular love and is neglected at our peril. Like a monster that needs feeding or a vital garden, our last garden, that needs tending. I am amazed at the strength of my desire to return, despite the ambiguous beauty of this hostile Eden.

It was a journey to another world without the galling necessity of death. The desolation and absoluteness of the 80th parallel and its neighbourhood wrestles with everything we carry around in our choice-drowned heads. Its pared-down world of clicking ice and sharp air, its spectral animals and light games, these are True.
To hell with Real.

*'I absolutely refuse to leave until we are able to take back with us something in the nature of a chart'*
Sir Arthur Conan Doyle, 'The Lost World'

⊕ **Overwhelmometer** Lenticular, 2004 / 80°05N, 19°05E

'THESE FRAGMENTS OF THE REAL WORLD, SLIVERS THAT EXPLAIN THE REAL WORLD, ARE ALSO FILTERING THROUGH TO HOW THE WHOLE IS CALIBRATED, UNDERSTOOD AND RECALLED.'

⊕ **The Trouble-We're-In-O'Clock** Lenticular, 2004 / 78°65N, 21°05E

◆ **Contextascope** Lenticular, 2004 / 79°65N, 21°10E

⊕ **David Buckland & Ian McEwan** 2005 / 78°30N, 16°35E

⬤ **Burning Ice** David Buckland, Director, Cape Farewell

Burning Ice, The Cold Library of Ice, Sadness Melts; texts projected on a glacier wall as the captain manoeuvres the Noorderlicht to within 5 metres of ice that has not been exposed to air for tens of thousands of years. It crumbles, crashing into the sea, carrying its history, soon to be melted away. A million years of the history of our planet is locked in ice two miles down. Once extracted, each tiny bubble releases air from the Earth's past, telling stories of temperature, $CO_2$ levels and the possibility of life. This ice we gently sail past, as dawn breaks on a cold morning, is the library of our past that now, with our irresponsible actions, we are causing to melt – Burning Ice.

The artists who have travelled as part of the Cape Farewell expeditions have told personal stories of change; they have made works on a human scale about what is a global problem. We have experienced the front line of climate change. In the High Arctic it is possible to witness just how fast the ice is melting and the balance of our planet is changing.

On Mars and Saturn the atmosphere is mostly $CO_2$, a lifeless gas in a world without life. It is perhaps life itself on Earth that has orchestrated a cocktail of gases – oxygen, nitrogen, hydrogen – which breed more life. The cycle of existence is held in biological balance. Into this blend we are releasing dangerous amounts of $CO_2$, fuelling our need for excess consumption. According to climate scientists, we are now 20-30 years away from a tipping point. From a point where more breeds more; the critical point of meltdown when the $CO_2$ in the atmosphere attracts so much heat that it becomes a self-propagating continuum, heat to more heat, life to less life.

I could cheer as we witness another 60,000-ton wall of ice crash into the sea, it is a spectacular and awe-inspiring event, but my emotions are not of wonder, but anger. How can we be so irresponsible, so wasteful with the lives of our children? It is so unnecessary; we now have the technological and economic skills to produce the energy we need without threatening the beauty of the place we inhabit, the beauty and the very possibility of life itself.

In the last 160 years we have dug up half the carbon resources nature has carefully hoarded. This process took $CO_2$ from the atmosphere, stored the excess and maintained the balance of the gases that breathed life into us, and the startling biodiversity that forms our unique planet. Now, we burn a ton of coal, extract a meagre 35% of its energy and release over three tons of $CO_2$ into the atmosphere. This is a very dirty exchange, unnecessary, ugly, and unsustainable.

The name 'Cape Farewell' plays with 'farewell', an expression of good wishes at parting, which here can evoke a sense of loss and finality, and the notion of a cape as a place of turning. We have choices to make. The artwork and texts in this book are about a way of imagining. Can we save a place in our imagination, alongside our capacity to conceive of ourselves, our needs and desires, our familial and societal relationships and responsibilities, for a larger vision? In our mind's eye, can we contain a pure biological compulsion towards life that can overcome our everyday struggles and balance our lives with the needs of the planet? At times of decision-making we must strive to preserve the possibility of our lives, it is in our own self-interest. This can only be achieved if we all engage in a collective system of sustainability. It is simple; at every turn reduce the need to pollute the air with $CO_2$ and convince others to do the same.

I quote the scientist Lynn Margulis in the conclusion to her book, 'The Symbiotic Planet: A New Look at Evolution':

*So far the only way in which we humans prove our dominance is by expansion. We remain brazen, crass and recent, even as we become more numerous. Our toughness is our delusion. Have we the intelligence and discipline to resist our tendency to grow without limit? The planet will not permit our populations to continue to expand. Runaway populations of bacteria, locusts, roaches, mice and grass always collapse. Their own wastes disgust, as crowding and severe shortages ensue. Diseases, as opportunistically expanding populations of the 'other', follow… We people are just like our planet mates. We cannot put an end to nature; we can only pose a threat to ourselves.*[1]

[1] p128 Lynn Margulis, 'Symbiotic Planet: A New Look at Evolution', Basic Books, Massachusetts, 1998

⊕ **Messenger** 2004 / 79°40N, 21°40E

'Think about the interaction of two particles, think of two lines meeting – a 'space-filling curve' – think of space being bigger than a line or a surface being bigger than a line but actually, in a sense, they have the same number of points.

There are the same number of points drawn on a plane as if drawn a line … this seems counter intuitive but both are actually infinite. A space-filling curve does crazy things … it's so bendy that it fills up the surface or volume completely.

If we do these surfaces with lines on them we can trace the lines in such a way that in the end all that would exist is the lines.

It would be a strange set of lines because they would have been infiltrated with the crumpled surface.'

David Buckland

⊕ **Northern Lights** 2004 / 71°10N, 20°15E

'As two lines drift towards the ambiguity of a crumpled surface, a situation develops that soon becomes irretrievable.'

### Here Today  Kathy Barber, Artist

In the vastness of the landscapes there is a real sense of the minute and monumental, permanence and flow. A continuous horizon contrasted by barely visible plankton and lichen, the pattern of wind traced in the surface of a frozen lake. This stark sense of scale and perspective is something you can easily become detached from in the city. The concept of a 20 million year timeframe actually seems to make sense in Svalbard. Here, you can see and feel evidence of the timescales that scientists work with, in the layers etched into glacier walls and the surfaces of ice caps. We were in an environment that has never had an indigenous population, a landscape that is remarkable for its relative lack of human intervention. The cities we've created around ourselves manage their distractions effectively to keep us focused on the present, the next minute, the next hour. A landscape like Svalbard puts this insistence on the here and now into context.

I wanted to create a conversation between these two different places, embedding small messages or signs that aim to bring something of that perspective back into an urban environment. They are small sketches, moments of quiet, something that might make you pause for a second. Signs for public spaces not intended to direct or sell. The neon and light boxes are solar-powered and self-contained. The light boxes are designed to blend into their surroundings during the day whilst they charge, automatically activating in low light. They have a life of their own, reacting to their surroundings and the daily patterns of sunlight.

**Light boxes** 2006 / 51°31'N, 0°01'W

##  People and Places at Risk  Prof. Diana Liverman, Director, Environmental Change Institute, Oxford

Climate scientists are producing ever more serious observations and predictions that tell us that the world is already warming and that future changes may be larger and more discontinuous than previously anticipated. Translating this climate science into meaningful impacts for landscapes and livelihoods is a complex and rather alarming task.

One image that is often used to synthesise ideas about the impacts of dangerous climate change is known as the 'burning embers' diagram adopted by the working group on impacts of the Intergovernmental Panel on Climate Change. The diagram links predictions of average global temperature increases to risks of dangerous changes to unique and threatened ecosystems, of extreme climate events, of unequally distributed impacts, of large overall impacts and of large scale discontinuities. The shading from lighter yellow to glowing deep red represents a shift from moderate to more severe risks.

Climate science warns us that, unless we dramatically reduce carbon emissions, we may well experience the more severe temperature changes and higher risks, and so we need to consider the impacts of these more serious changes on ecosystems and humanity.

Warming beyond two degrees will bring the widespread collapse of arctic ecosystems as ice dependent polar bears and seals lose habitat and the boreal forest invades the tundra. In the tropics, species that live in the cooler highlands, such as many birds and amphibians, will be squeezed upwards, bringing them into conflict with other species and risking extinction as their habitat shrinks or disappears from the drier or warmer mountaintops. Rising sea levels and changing ocean chemistry will threaten coral reefs, drown mangroves and estuaries, and increase risks from tropical storms. Those people who depend on these ecosystems are also at risk. Arctic, Alpine and coastal peoples are losing the ecological basis of their cultures. For low lying islands such as Tuvalu or Kiribati in the Pacific, global warming could mean the abandonment of their viable territory through sea level rise, saline intrusion into water resources, and higher risks from tropical storms. If temperature increases by more than two degrees, one estimate places more than two billion people at risk from water shortages, 200 million more people at risk from malaria and 50 million at risk from hunger and from coastal flooding.

But biophysical vulnerability is only one element in surviving climate change, where impacts often depend on the social and economic characteristics of affected peoples and on the intersection of climate change with other pressures on ecosystems. The risks of sea level rise and severe storms are vastly compounded when millions of poor people live along low lying coasts and will have to adapt, migrate or perish en masse as a result of climate change. The risks of species extinction are higher when land use prevents a shift in range, or harvesting places extra pressure on populations at risk. Even more complex are the interactions between climate risks and other stresses to the food supply. Millions may be at risk from changing climate, land degradation and poverty in Africa.

For many people the onset of global environmental change is coincident with the spread of economic globalisation. The suite of policies of free trade, less government and resource privatisation are labelled variously as 'neoliberalism', 'market environmentalism' or the 'Washington consensus'. This means that rural producers are doubly exposed to climate change and globalisation, providing opportunities for some, but risks to the many that are losing the productivity of their land and government support at the same time, as they must pay for privatised water and compete in the international market. I have visited farmers in communities in Mexico, where free trade and neoliberalism have brought few benefits. They are facing warmer temperatures and more intense storms without agricultural extension support, and are in debt, with declining prices for their products whilst they pay high costs for water, fertiliser and food.

Global warming is occurring in a highly unequal world and contributes to further inequality because responsibility for carbon emissions, the geography of vulnerability and impacts, and actions to respond to the situation are differentially attributed within and between countries. Thus, the United States, responsible for more than a third of emissions and high per capita carbon use, has been unwilling to join the Kyoto protocol. Countries such as Bangladesh, with modest emissions, especially on a per capita basis, are forecast to experience devastating impacts from a warming world because of the physical vulnerability of low elevations and the social vulnerability of a poor and expanding population. Even in the United States, Hurricane Katrina showed us how growing inequality within the wealthiest countries can create widespread vulnerability to the climate and immense social disruption.

Climate change is going to affect the basic needs of large numbers of people. It will threaten the security of their food, water, and health and prompt them to move, rebel or even sue those who they think are responsible for their suffering. Without heroic efforts to reduce carbon emissions we will have to focus much of our efforts on helping the world to adapt to climate change.

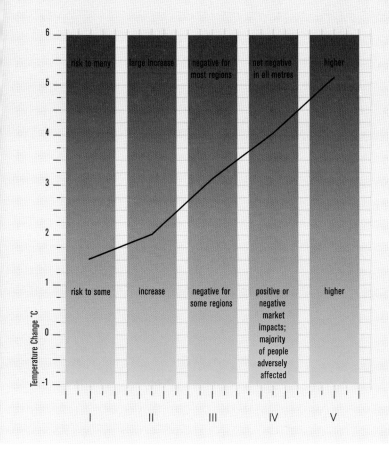
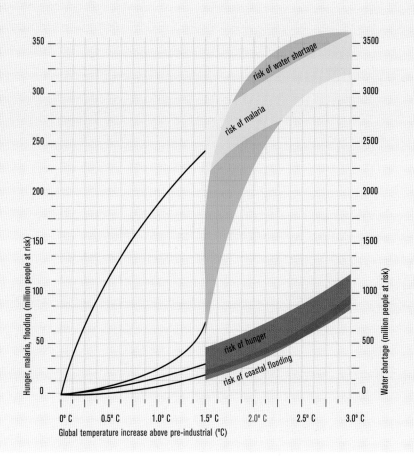

▲ **Burning embers**
  I   Risk to Unique and Threatened Systems
  II  Risks from Extreme Climate Events
  III Distribution of Impacts
  IV  Aggregate Impacts
  V   Risks from Future Large-Scale Discontinuities

Probabilistic Integrated Assessment of 'Dangerous' Climate Change
Michael D. Mastrandrea and Stephen H. Schneider
Science 23 April 2004: Vol. 304. no. 5670, pp. 571 – 575

▲ **Millions at risk**
Parry, M.L., Arnell, N.W., McMichael, T., Nicholls, R., Martens, P., Kovats, S., Livermore, M., Rosenzweig, C., Iglesias, A. and Fischer, G. Millions at risk: defining critical climate change threats and targets. Global Environmental Change, 11, (3), 2001, pp.181-183.

⊕ **80° North** 81° 30N, 16° 30E

'THE LIGHT IS PURE UNADULTERATED PHOTON, SOMETIMES HAVING THE DISTURBING EFFECT OF LOOKING LIKE A COMPUTER SIMULATION – TOO CRISP, TOO CLEAR.'

154 David Buckland

⊕ **Waterwall** 2005 / 51°33N, 2°35W

'WE'VE GOT OURSELVES INTO THE SITUATION WHERE WE ARE HAVING TO ADDRESS THE NEEDS OF PEOPLE UNBORN. AND NOT OUR OWN CHILDREN. MAYBE NOT EVEN OUR CHILDREN'S CHILDREN BUT GENERATIONS AFTER THAT. EVEN THE MOST IDEALISTIC OF THINKERS AND ACTORS ON THE WORLD STAGE IN THE PAST HAVE ONLY REALLY ADDRESSED THEMSELVES TO PROBLEMS IN THE PRESENT.'

IAN MCEWAN

David Buckland 155

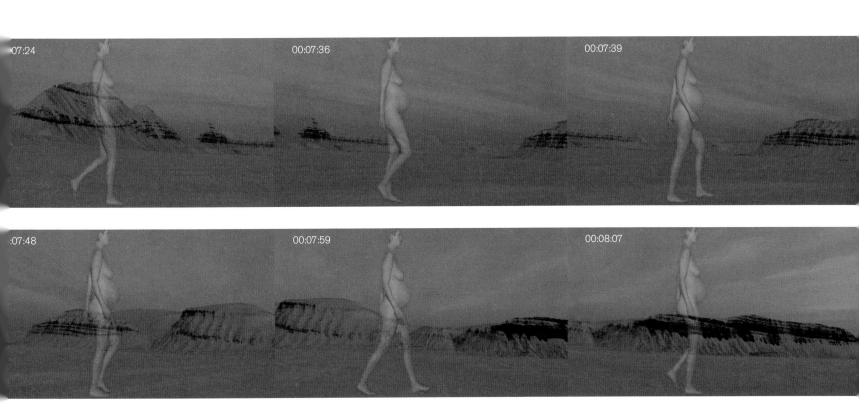

⊕ **The Ice Garden, Bodleian Library, Oxford** 2005 / 51°33N, 2°35W

## Climate Change and Injustice
Dr. Saleemul Huq, Director, Climate Change Programme, International Institute for Environment & Development

The problem of human-induced climate change is widely recognised as having been created primarily because of greenhouse gas emissions from rich countries since the Industrial Revolution. Even in more recent years, major emissions continue to be associated with the activities and the consumption patterns of the richer citizens of the globe, located mostly in industrialised countries but also, increasingly, in poorer countries. The adverse impacts of climate change, however, will primarily be felt by the poorer citizens of the world, located mostly in the poorer countries but also in the richer ones. This mismatch between the citizens causing the problem and those who will suffer the consequences has not yet been adequately addressed. The citizens of the world must acknowledge this situation and let it guide their individual actions with respect to the problem.

The problem of climate change is the quintessential global problem in that it is caused by the actions of every single citizen and is likely to then affect each citizen in turn, to varying degrees. Hence the solution needs to include individual action on the part of each and every citizen. Those that are emitting more than their fair share of greenhouse gases should consider ways of both reducing their emissions and providing compensation for those who emit very little but will suffer the consequences of emissions, such as most of the poorer populations in Africa and parts of Asia and Latin America. The kinds of actions that can be taken are 'mitigation', to reduce emissions and 'adaptation', to enhance resilience against the adverse impacts of emissions. Those who emit more, which is most of the citizens of industrialised countries, need to both reduce their emissions and contribute to poorer countries and their most vulnerable communities. They must help them to adapt to the inevitable consequences of climate change, by providing financial assistance, for example.

One way for the aware and concerned citizen to act is to calculate his or her emissions or 'carbon footprint' and then try to reduce it as much as possible by reducing those actions that cause most emissions. As it will not be possible for anyone to reduce their carbon footprint entirely, they can then pay for 'carbon offsets', now readily available through the carbon market. They would pay to reduce the equivalent amount of carbon to the residual emissions that they were individually responsible for.

Calculating one's carbon footprint and making payments for carbon offsets is now relatively easy but only addresses the adverse 'environmental' impacts of emissions. It does not alter the consequences of one's impact on people, mostly the poor, which will still happen, and cannot be 'offset' by buying carbon reductions alone. Hence the truly worried citizen must also show concern for the poor and vulnerable humans who will be affected by the adverse impacts of climate change and not only the adverse impacts on the 'environment'. There are not yet ways for citizens to make such compensatory payments directly but it is certainly time to think of them.

**Glacial Climb** Alex Hartley 2004 / 78°55N, 12°00E

160  Education

IT HAS BEEN CAPE FAREWELL'S AMBITION TO INFORM YOUNG PEOPLE ABOUT CLIMATE CHANGE, IT IS THEY WHO WILL BE MOST AFFECTED BY IT.

### Cape Farewell Education David Buckland, Director, Cape Farewell

From the outset Cape Farewell's ambition has been to inform young people about climate change, it is they who will be most affected by it. Climate change is a complex subject that spans several disciplines. It is also a very exciting field to be working in at the moment and this has motivated our educational projects. We have built education modules which are now in use in schools throughout the UK and we continue to develop our website to make this resource material more widely available. Our major partners in this endeavour are the National Oceanography Centre, Southampton, Big Heart Media directed by Colin Izod, the Geographical Association, the Nuffield Foundation and we have a funding partnership with NESTA.

In 2004 the 'Extreme Environments' module of the Geography 21 GCSE was built from work done on the first Cape Farewell expedition. Fred Martin from the Geographical Association created a 180-page teachers' and students' manual with Garry Doyland. There is an accompanying 40-minute film directed by Colin Izod, an interactive CD-rom and a student resource book with colour illustrations. In 2005, science teachers Suba Subramaniam and Mike Vingoe, working closely with NOC and Big Heart made an educational film and wrote science oceanography modules dealing directly with climate change. The work is currently being developed in partnership with Andrew Hunt and the Nuffield Foundation and the first module, on plankton, is now part of the Science 21 GCSE for 14-16 year olds.

Visit www.capefarewell.com

**Garry Doyland** – *Geography Teacher*
Glaciers have been likened to mighty rivers of ice. It's hard to imagine, but the different kinds of movement that you see in a river — eddies, rapids, waterfalls — all occur in a glacier, only many times more slowly. Glaciers have similar changes in flow rate and often form falls of fast-moving ice above slow-moving ice pools. Glaciers flow faster down their centres than at the edges, or ice margins, and more quickly at the surface than at the bottom or bed.

**Mike Vingoe** – *Science Teacher*
I think my purpose on the Cape Farewell expedition was to bring the professional scientists down to a level that I know fifteen and sixteen year olds would be able to comprehend, and to look at some of the things they were doing and try to develop experiments that I think ten or eleven year olds could understand. So that hopefully what we bring back is things that children can do in the lab which help them to understand that global warming is happening. Are we going to give them a choice? Do we care about it? Should we take action? What are we going to do about it? Or are polar bears really not that important?

**Suba Subramaniam** – *Science Teacher*
Biology being my specialism within science teaching, being in the Arctic environment amongst the wildlife: getting close to the walruses, listening to the sounds of the bearded seals, seeing polar bears, looking at the birds and the whales was just the most amazing experience. I was actually in amongst these creatures I teach about all the time, when we study food webs, food chains and how they are affected by the environment. One of the first things I teach in the environment unit is how polar bears are adapted to live in their surroundings.

**Journal, 2005** Rachel Whiteread, Artist

Having frozen and thawed, frozen and thawed, frozen and thawed, I'm beginning to get the hang of how many layers of clothes to wear.

My main objective was to walk. I wanted to witness the whiteness of it all. The first moment I came out, I felt like an astronaut – like we'd just landed on this deserted place. You feel you shouldn't actually be here. It is not a place for humans; it's such a hostile environment and if you are not careful, the cold will just take your life force in seconds.

I've now been on two completely sublime walks where I have felt totally humbled by the environment. The scale of the place, the 360° vistas, the different tones of white and the longest blue shadows I've ever seen, are all simply breathtaking. The silence, when you can actually experience it, is deafening.

I've concentrated on two main activities when walking. Firstly the sound and feel of one's footsteps (silent, soft, loud, crunching and resonating – your tread on the different densities of snow and ice making music). Secondly, searching for signs of life; polar bear and Arctic fox tracks, Svalbard ptarmigans (that are apparently so stupid they willingly stand by and watch their friends being shot), Svalbard reindeer (Svalbard pigs as the locals fondly call them), various lichen (so colourful, robust and beautiful in this environment), blades of grass, and finally the odd dried leaf. Oh yes, and of course the very noisy homo sapiens on their skidoos.

Rachel Whiteread 2005 / 78°30N, 16°35E

Glacial Valley 2005 / 78°30N, 16°30E

**⊕ Embankment** Tate Modern 2005 / 51°52N, 1°95W

'ONE OF THE THINGS I HAVE DREAMT ABOUT DOING IN THE PAST WAS CASTING A VALLEY. I THINK, IN A WAY, THAT IS WHAT I HAVE DONE. MAYBE THAT IS WHAT IT IS – A CAST OF A GLACIAL PLAIN.'

# Turning Points
Robert MacFarlane, Novelist

One day early last March, I walked over miles of rough and sunlit ground to Cape Wrath, the most northwesterly point of the British mainland. Cape Wrath, like many place-names in the west and north of Scotland, is from the Old Norse, and 'Wrath' means not 'anger', as might be assumed, but 'turning-point'. It was when Viking raiders, returning from long Atlantic voyages of exploration and plunder, rounded the Cape's distinctive red sandstone cliffs, that they knew they were heading home. And on the way out to raid, the Cape marked the point beyond which they knew home had been left far behind. For the Norsemen, it was a pivot in the ocean world, beyond which lay either fight, or respite.

From the miles I walked during that day, I remember a buzzard's shadow cast on the land, flicking up and down as it passed over undulating ground. I remember the sea, docile and silvered, always to my left. I remember gulls planing the wind in wide white curves, and turning on their wing-tips. I remember, too, stopping at a nameless stream-cut, in whose water tiny black trout sped and darted, and picking up a bleached gull skull. When I rotated the skull in my hands, I could hear the muted run of sand grains through its chambers. At the same stream-cut, I found and kept a single white stone, a soft sphere of quartz, round as an irisless eye, for it was smooth and lovely to handle.

As I walked that day, I was more than usually aware of the intent of the land and the sea, of their ongoing and vigorous concerns. The silver sea, the red rocks, the brown moor. Aware too of their indifference to my presence. Near the Cape itself, I stopped on a high promontory, which looked northwest into the bright extensive air. I drank cold water from a cup, and watched the red headland of the Cape rising nearly four hundred feet out of the sea, and the sixty feet of white lighthouse reaching above it. I could see far out to sea. The dark horizon line was as clear as a strap. I remember now that, as I sat there on the headland, my mind drifted out over miles of imagined ocean, to another northwesterly place. To Banks Island, in the Canadian High Arctic, home of many of the Inuvialuit people.

Banks is one of the places in the world in which the effects of climate change are now most apparent. On Banks, indeed, the consequences of climate change are perceptible in language as well as in degrees Celsius. There, environmental shifts are happening so fast that the Inuvialuit inhabitants do not have the words to describe what they now see around them. New species of fish, bird, and insect have migrated north to the island, following the isobars. Autumn thunder and lightning have been witnessed from Banks for the first time. 'Permafrost' is no longer tolerable as a term, for the ground-ice is melting: in Sachs Harbour, the main settlement on Banks, buildings are subsiding and road surfaces are slushing up.

There have been disappearances as well as arrivals on Banks. The permafrost melt has caused an inland lake to drain into the sea. The intricate stages of hardening through which the sea-ice around Banks cycles – frazil, grease, nilas, gray – are no longer being fulfilled in many places during summer, for the temperature of the sea water is spiking above the key freeze-point of 28.6°F. The Inuvialuit culture is unprepared for these rapid fluxes. Old words (the name of an inland lake) are now unaccompanied by their phenomena; new phenomena (a fork of flame in a previously lightningless sky) are unaccompanied by words.

Places like Banks Island are the barometers of climate change. Changes that are imperceptible in Scotland, or Iowa, or Cologne, express themselves very visibly in the globe's high latitudes, and its high altitudes. Such places presently act as warnings to us. Unless we change direction, however, such places will stand as elegies, memorials.

For we have arrived at a turning point, perhaps the most significant in the history of humanity. It is now clear that there is no more urgent intellectual task facing the human species – that there has never been a more urgent task facing the human species – than thoroughly to re-imagine its relationship with nature. We need to learn to find in our environment not something to be despised or pressed always into service, but something instead to be wondered at, to be revered, to be understood on a basis of mutuality.

In a decade's time, we will look back at these two or three years as a pivot point of human history. The potential gains if we turn the right way are immense; the potential losses if we turn the wrong way, incalculable. And what is required to make the correct turn is nothing less than a reimagining of the terms by which we understand happiness, wealth, progress, culture, and even time. We need to revive an older form of human awareness, according to which nature is neither backdrop nor storehouse, neither product nor chattel, but a community of which we are inevitably a part. It has become of compelling need that we move towards a respectful cherishing of the natural world, rather than aspiring to its complete control. The American ecologist Aldo Leopold saw this fifty years ago, before climate change had even been detected. 'We abuse the land because we regard it as a commodity belonging to us,' he wrote. 'When we see land as a community to which we belong, we may begin to use it with love and respect'.

That day, I turned from Cape Wrath, and walked back south for hours, through the remarkable light which was falling upon the land from the northwest. On the Atlantic coasts of the British Isles, the air possesses a remarkable transparency, for it is almost free of particulate matter. Little loose dust rises from the wet land, and the winds blow prevailingly off the sea. Through such air, the photons can proceed without obstacle. The light moves, unscattered, and settles upon the forms and objects of those regions with an unconcealing candour. Standing within such a light, one feels thankful for the light's openness, and one recalls and understands the ancient and enduring association within theology of light and grace. There is a sense of something having being freely given, without its store having being diminished. Walking within that light that day, I thought gratefully that the memory of this sunlight would last me for years to come.

Cape Farewell celebrates the commitment, hard work and inspiration of the artists involved, without which none of this endeavour would have been possible. Over the past 50 years many scientists have been working to establish just how serious our concerns over climate change should be. The scientists involved with the Cape Farewell project have been generous with our requests and excited by the possibility of working with artists.

As well as those who have contributed to the book and been part of the expeditions, David Buckland, Cape Farewell would like to thank:

**Cape Farewell Board of Governors**
Graham Devlin, John Hammond, Charlie Kronick, Fiona Morris, Michael Wilson

**Cape Farewell HQ**
Sam Collins, Karen Fardell, Emma Gladstone, Greg Hilty, Anita Ingram, Alexandra Lambert, Vicki Lewis, Vicky Long, Sarah Macnee, Janette Scott

**Cape Farewell Film**
Directors: David Hinton & Colin Izod
Camermen: Nick Edwards 16mm, Philip Chavannes, Ole Bratt Birkeland
Sound: Albert Bailey, John Burns
Editor: Duncan Harris
Runners: Joe Chapman, Sean Buckland, Andy Symon

**Noorderlicht**
First Mate Maaike Groeneveld, Captain Ted Van Broeckhuysen, Captain Gert Ritzema and the ship's cooks

**Funders**
Arts Council England, The Bromley Trust, Calouste Gulbenkian Foundation, Lighthouse Foundation, National Oceanography Centre, NESTA, The Nuffield Foundation

**And**
Judy Adam, Sian Alexander, Artangel, Bergit Arends, Philippa Barr, Dr Bob Bloomfield, Bolton & Quinn, Catherine Boyd, Dr Barry Buckland, Denis Buckland, Piera Buckland, Anne Canning, Colin Challen MP, Caryl Churchill, Michaela Crimmin, Ian Curtis, George Dub, Eden Project, Siân Ede, Kim Evans, Richard Fielden, Bronac Ferran, Sarah Fletcher, Tish Francis, Ed Gillespie – Futerra, Peter Gingold, Teresa Gleadowe, Salette Gressett, Fam van de Heyning, Sue Hoyle, Andrew Hunt, Deborah Jackson, Cathy James, Jude Kelly, John Kilroy, Sir David King, Dr Ko de Korte, Dr Charles Kriel, David Lambert, James Lovelock, Mark Lynas, Scott Martin, Professor Jochem Marotzke, Fred Martin, Deborah May, Malcolm McCulloch, Victoria Miro, Paul Morrell, Michael Morris, Charlotte Mullins, Bill Paterson, Dr David Patterson, Sir Hayden Phillips, Cornelia Parker, Alison Purvis, Philip Pullman, Lord Puttnam, Professor Chris Rapley – British Antarctic Survey, Jonathan Saunders, Deborah Saxon, Julia Simon, Sarah Warsop, Jane Wentworth, Admiral Sir Alan West, Dr Chris West, Dr Richard Wood – Hadley Centre

**Photography**
Heather Ackroyd & Dan Harvey: 2, 4, 83, 102, 105, 106, 108, 111, 115, 116, 117, 118, 119, 136, 160, 171

Badger: 33, 58, 156

Kathy Barber: 10, 26, 48, 143, 145

Gautier Deblonde: 61, 62, 63, 64, 65, 66, 67, 68, 69, 70, 71, 78, 87, 88, 109, 120, 135, 166, 169

David Buckland: cover, back cover, 1, 12, 20, 28, 31, 36, 37, 40, 52, 56, 94, 97, 98, 99, 112, 123, 124, 126, 134, 138, 141, 150, 154, 158, 163, 164, 176

Nathan Gallagher: 144, 145

Antony Gormley: 38, 39, 172

Marije de Haas: 32, 59

Alex Hartley: 19, 84, 146

Vicky Long: 92

**2003 Expedition**
Albert Bailey Soundman
David Buckland Artist
Valborg Byfield Oceanographer
Philip Chavannes Cameraman
Gautier Deblonde Photographer
Garry Doyland Geography Teacher
Max Eastley Sound Artist
Nick Edwards Artist
Gretel Ehrlich Novelist
Sarah Fletcher Oceanographer
Dan Harvey Artist
Caspar Henderson Journalist
David Hinton Film Director
Gary Hume Artist
Colin Izod Film Director
Liam Izod Sat-com
Alexandra Lambert Project Manager
Carolyn Maze Sailor
Suba Subramaniam Science Teacher
Andy Symon Runner

**2004 Expedition**
Heather Ackroyd Artist
Albert Bailey Soundman
Kathy Barber Artist
Emily Boxall Student
Simon Boxall Oceanographer
David Buckland Artist
Sean Buckland Runner
Joe Chapman 2nd Camera
Philip Chavannes Cameraman
Quentin Cooper Radio Journalist
Gautier Deblonde Photographer
Max Eastley Sound Artist
Nick Edwards Artist
Sarah Fletcher Oceanographer
Alex Hartley Artist
Dan Harvey Artist
David Hinton Film Director
Colin Izod Film Director
Michèle Noach Artist
Mike Vingoe Science Teacher

**2005 Expedition**
Heather Ackroyd Artist
Ole Bratt Birkeland Cameraman
David Buckland Artist
John Burns Soundman
Juliet Butler Journalist
Peter Clegg Architect
Siobhan Davies Choreographer
Gautier Deblonde Photographer
Max Eastley Sound Artist
Nick Edwards Artist
Antony Gormley Artist
Duncan Harris 2nd Camera
Alex Hartley Artist
Dan Harvey Artist
David Hinton Film Director
Charlie Kronick Environmentalist
Ian McEwan Novelist
Andy Robbins BBC Culture Show
Dr Tom Wakeford Ecologist
Rachel Whiteread Artist

**⊕ 2005 Crew**

Max Eastley
Charlie Kronick
Nick Edwards
Alex Hartley
Antony Gormley
Peter Clegg
Siobhan Davies
Rachel Whiteread
David Buckland
Gautier Deblonde
Dr Tom Wakeford
Heather Ackroyd
Dan Harvey
Ian McEwan